NEW RESEARCH ON ENERGY ECONOMICS

PAUL G. CALDWELL
AND
EARL V. TAYLOR
EDITORS

Nova Science Publishers, Inc.
New York

NEW RESEARCH ON ENERGY ECONOMICS

For permission to use material from this book please contact us:
Telephone 631-231-7269; Fax 631-231-8175
Web Site: http://www.novapublishers.com

LIBRARY OF CONGRESS CATALOGING-IN-PUBLICATION DATA

New research on energy economics / Paul G. Caldwell and Earl V. Taylor, (editors).
p. cm.
ISBN 978-1-60456-354-2 (hbk.)
1. Power resources--Economic aspects. 2. Energy development--Economic aspects. I. Caldwell, Paul G. II. Taylor, Earl V.
HD9502.A2N494 2008
333.79--dc22 2008003388

Published by Nova Science Publishers, Inc. ✦ New York

CONTENTS

PREFACE

This new book presents the latest research from around the globe in the field of energy economics which is a subfield of economics that focuses on energy relationships as the foundation of all other relationships. It is a subfield of ecological economics in that it assumes that food chains in ecology are directly analogous to energy supply chains in human industries. Coverage includes issues related to optimal production and consumption affecting energy, minerals, land, air and water, and includes analysis of firm and industry behavior, environmental issues and public policies.

In the analysis and design of energy systems, techniques are often used which combine scientific disciplines (mainly thermodynamics) with economic disciplines (mainly cost accounting) to achieve optimum designs. For energy conversion devices, cost accounting conventionally considers unit costs based on energy. Many researchers have recommended that costs are better distributed among outputs if cost accounting is based on the thermodynamic quantity exergy. One rationale for this statement is that exergy, but not energy, is often a consistent measure of economic value. Several exergy-based economic-analysis methodologies exist.

In Chapter 1, the relations between exergy and economics are discussed, and many existing analysis techniques which integrate exergy and economics are reviewed. Some of the exergy-based economic-analysis methodologies considered include exergoeconomics and thermoeconomics, exergy-based pricing, EXCEM analysis and analysis based on the ratio of thermodynamic loss to capital cost. The latter approach identifies as important the ratio of thermodynamic loss rate to capital cost. In that approach, data for devices in systems are examined, and correlations are identified between capital costs and specific second-law-based thermodynamic losses (i.e., total and internal exergy losses). Such correlations can improve understanding and aid design efforts. The goals of most such techniques include the determination of the appropriate allocation of economic resources for optimal design and/or operation, and/or the economic feasibility and profitability of a system. Several illustrative examples are considered.

Food of local origin is often said to be environmentally friendly because of small transport distances. In public opinion, long distance transportation is equated with high energy turnover, without asking for the means of transportation. But the specific energy turnover decreases with increasing size of the transportation unit. One commuter bus requires much less energy compared with 40 cars and one HGV is more effective than 100 light trucks. Looking at ecological impacts, the energy turnover is one of the most important matters, and is closely related to greenhouse impact also.

From economics we know, that the specific financial effort depends upon the number of items which are produced, because of the fixed costs to be taken into account. This well-known economics law is called "The Economy of Scale". Regarding the ecology, we actually assume the same regularity as in economics: Above all, the specific ecological impact should be dependent on the number of produced items. They assume small business units to be less effective in terms of ecology compared with bigger units. And it should be possible to derive a minimum business size as a "break even" area. This central hypothesis they call "The Ecology of Scale".

To investigate the hypothesis in the food business, case studies with juices, lamb meat, beef, pork, wine, and apples have been performed. In all cases food of local origin is compared to global food. In the first step, the energy turnover of the entire process chain has been investigated, including farming, crop and breed, transports of raw, intermediate, and final products as well as the efforts of distribution up to the point of sale. All energy efforts are allocated to the functional units.

The results in Chapter 2 demonstrate a declining relation of the specific energy turnover and the business size. Small units require more energy per food item compared with bigger enterprises, independent of local or global origin. Surprisingly, the authors can even find local farmers, which can cause higher specific energy turnover compared with some global process chains, because of high sophisticated logistics and good energy saving strategies of bigger units. But – on the other hand – regional food can compete with global business in terms of energy in the case of sufficient business size. Additionally, the researched data indicate a minimum business size as break even. All these results support the theory of Ecology of Scale. In all case studies, the authors find regional food business meeting these requirements as well. But, there is no indication to blame global food as energy wasting process chains.

The Asia Pacific region of the world is home to some of the fastest growing economies in the global economy. The vast majority of countries in the Asia Pacific region are net importers of crude oil and rising oil prices could impact economic and financial activity. Chapter 3 uses a multifactor market model to investigate the impact that oil price changes have on the equity markets of the Asia Pacific economies. Results are presented to show that oil price movements are an important factor influencing the stock returns for approximately half of the countries studied. Asymmetric oil price changes are important factors influencing stock markets for the countries of India, Japan, South Korea, and Sri Lanka. Asymmetric oil price volatility changes are important determinants of stock prices only in the case of Singapore. Managers, investors, and policy makers should find these results useful.

Existing literature on environmental economics has mainly focused on comparative analysis of environmental policy instruments such as direct regulations, emission taxes and emissions trading. In the literature, market-based approaches are recommended as policy instruments for combating climate change. However, new technologies must be created for the long-term solution to climate change problem. Although environmental policy can treat external diseconomy, market failure associated with research and development (R&D) activities must be addressed by technology policy: the authors are faced with the double market failure. Therefore, the integration of environmental and technology policies is needed for combating climate change. Japan's experience of pollution control in the 1960s and 1970s provides a case of such policy integration. Chapter 4 investigates whether Japan's policy responses to air pollution during the period were effective. Recently, in Japan, a new approach was adopted in order to enhance the energy efficiency of appliances and vehicles. This is called the Top Runner Program, which may be able to promote R&D competition among producers. The chapter also discusses the effectiveness of the program.

Chapter 5 briefly examines the future prospects for the economical viability of large-scale renewable energy production using maricultured cyanobacteria. In order to reduce CO_2 emissions from burning fossil fuels in appreciable amounts, the replacement energy source will by necessity be substantial in scale. Solar energy is the most likely candidate because the amount of solar energy received on the Earth's surface is vast and exceeds the anthropogenic primary energy use by more than 6,000 times. Although solar energy is abundant, its economical utilization is not straightforward because the intensity received on the surface of the earth is relatively low. Current research and development efforts are focused on the production of biofuels as renewable, economical feasible energy sources from the land biomass. The authors propose, however, for reasons of scale and to minimize further environmental harm, that the utilization of the sea surface is a more viable alternative to land biomass exploitation. The sea surface area available for energy production far exceeds available cropland and use of the sea will not take valuable cropland out of food production. The current R and D strategy utilizes photosynthesis and the nitrogenase enzyme of cyanobacteria. The biological basis of relevant energy metabolism in cyanobacteria is briefly described. A model for future H_2 production systems is presented, and a very rough trial calculation of the cost of photobiological H_2 production is made in the hope that it may help the readers recognize the possibilities of large-scale H_2 production and understand the need for the research and development.

Using a multivariate vector-autoregression (VAR) approach, Chapter 6 studies the dynamic relationship among oil prices, real economic activity, real stock prices and real interest rates in Greece over the period 1989:1 to 2007:6. The empirical evidence suggests that oil price changes affect real economic activity and are important in explaining stock price movements. Employing vector error-correction models and generalized impulse response functions, the empirical results support the endogeneity of real economic activity and real stock prices however the signal transmitted from one variable to the other is weak.

In Chapter 7 the authors analyze the economic impact of oil prices in the G7 economies over the period 1957Q1 to 2007Q1. Using a variety of definitions for oil price movements, the results suggest the following. First, oil prices have a greater impact on inflation and economic growth rates when they are measured in local currencies and in real terms. Second they find evidence of an asymmetric effect of oil prices on both inflation and economic growth rates. Furthermore, the economic impact of oil prices is significantly lower for the period after the mid-1980s than for the first subsample. Finally, the economic impact of this variable differs for the countries analyzed in this chapter.

As explained in Chapter 8, the high price of gasoline was an important consideration during the debate on major energy legislation, which ended August 8 as the President signed the Energy Policy Act of 2005, H.R. 6 (P.L. 109-58). However, prices continued to surge, spiking at the end of August when Hurricane Katrina shut down refining operations in the Gulf of Mexico. The continuing crisis renewed attention to some issues that were dropped or compromised in the debate over P.L. 109-58, as well as a number of initiatives to reduce the impact of high prices on consumers.

A large number of factors combined to put pressure on gasoline prices, including increased world demand for crude oil and U.S. refinery capacity inadequate to supply gasoline to a recovering national economy. The war and continued violence in Iraq added uncertainty and a threat of supply disruption that added pressure particularly to the commodity futures markets.

Among the issues that received new attention were vehicle fuel economy standards, leasing on the Outer Continental Shelf, and refinery "revitalization" provisions. The Gasoline for America's Security Act of 2005, H.R. 3893, was passed by the House on October 7 by a vote of 212-210. A similar bill, S. 1772, was defeated in the Environment and Public Works Committee on October 26 by a tie vote of 9-9.

The budget reconciliation process was the vehicle for two major energy initiatives: the opening of part of the Arctic National Wildlife Refuge (ANWR) to oil and gas development and the lifting of the moratorium on oil and gas leasing on much of the Outer Continental Shelf (OCS). Both the Senate Energy and Natural Resources Committee and the House Resources Committee included leasing ANWR in their reconciliation bill sections. The House Resources Committee included provisions regarding OCS leasing, but the Senate Energy Committee did not. ANWR leasing was included in the Senate bill that passed on November 3 (S. 1932), but the House leadership dropped both ANWR and OCS provisions before its reconciliation bill, H.R. 4241, was passed on November 18 by a vote of 217-215. The conference report on the bill, approved by the House but amended in the Senate, does not contain either measure.

The gasoline price surge influenced the debate over P.L. 109-58, but the urgency of previous energy crises was lacking. In part, this may be due to the fact that there has been no physical shortage of gasoline or lines at the pump. In addition, the expectation of former crises — that prices were destined to grow ever higher — has not been prevalent.

However, the persistence of high gasoline and oil prices into a second summer has raised alarms over the economic consequences of the situation, heightened following the disastrous effects of Hurricane Katrina.

As gasoline prices surged to the $3.00 per gallon level again in spring 2006, a new wave of legislative proposals was put forth in Congress, including strengthening measures regarding price gouging and profiteering, temporarily refunding federal gasoline taxes, and windfall profits tax measures.

A growing consciousness of the intertwined issues of environment and development and the particular role of energy in the nexus of these issues have been realised at present. As a result of this awareness, voices of concern in support of Environmentally Sound and Sustainable Development (ESSD) are now increasingly heard. The great challenge now is to move from the voicing of concerns to accelerated action and change. The list of environmental issues that are relevant to any discussion of ESSD strategies is long and diverse. Implementing development strategies which include the alleviation of poverty as a central focus must address both the direct and indirect linkages between these environmental issues and poverty, as explained in Chapter 9.

In: New Research on Energy Economics
Editors: P.G. Caldwell and E.V. Taylor, pp. 1-66
ISBN: 978-1-60456-354-2

Chapter 1

INTEGRATING EXERGY AND ECONOMICS: AN ENHANCED APPROACH TO ENERGY ECONOMICS

*Marc A. Rosen**
Faculty of Engineering and Applied Science, University of Ontario Institute of Technology, 2000 Simcoe Street North, Oshawa, Ontario, L1H 7K4, Canada

Abstract

In the analysis and design of energy systems, techniques are often used which combine scientific disciplines (mainly thermodynamics) with economic disciplines (mainly cost accounting) to achieve optimum designs. For energy conversion devices, cost accounting conventionally considers unit costs based on energy. Many researchers have recommended that costs are better distributed among outputs if cost accounting is based on the thermodynamic quantity exergy. One rationale for this statement is that exergy, but not energy, is often a consistent measure of economic value. Several exergy-based economic-analysis methodologies exist.

In this chapter, the relations between exergy and economics are discussed, and many existing analysis techniques which integrate exergy and economics are reviewed. Some of the exergy-based economic-analysis methodologies considered include exergoeconomics and thermoeconomics, exergy-based pricing, EXCEM analysis and analysis based on the ratio of thermodynamic loss to capital cost. The latter approach identifies as important the ratio of thermodynamic loss rate to capital cost. In that approach, data for devices in systems are examined, and correlations are identified between capital costs and specific second-law-based thermodynamic losses (i.e., total and internal exergy losses). Such correlations can improve understanding and aid design efforts. The goals of most such techniques include the determination of the appropriate allocation of economic resources for optimal design and/or operation, and/or the economic feasibility and profitability of a system. Several illustrative examples are considered.

* E-mail address: marc.rosen@uoit.ca, Tel: 905/721-8668, Fax: 905/721-3370.

Nomenclature

CV coefficient of variation
Ex exergy
K capital cost
L thermodynamic loss
P pressure
Q heat
R ratio of thermodynamic loss rate to capital cost
SD standard deviation
T temperature

Greek Letters
η energy efficiency
ψ exergy efficiency

Subscripts
en energy loss
ex exergy loss
ex-i internal exergy loss
ex-e external exergy loss
m mean
max maximum
min minimum
o reference-environment state
od outdoor
re return
s station
su supply

Superscripts
* appropriate

1. Introduction

The merit of a system or process is usually based on a range of factors. Conventional parameters taken into account normally include technical performance and efficiency, economic viability, health and safety implications, environmental impact, resource scarcity and social acceptance. The merit of a system or process is best evaluated with methodologies that take into account the above factors and others. A systems viewpoint is required for comprehensiveness.

Economics is clearly central to the analysis and design of energy-conversion and energy-intensive technologies. Analysis, design and optimization activities for such technologies often utilize techniques that combine technical disciplines like thermodynamics with economic disciplines like cost accounting. Cost accounting applied to energy-related systems normally considers unit costs based on energy.

Many authors suggest that the first and second laws of thermodynamics have significant implications for economic theory. Bryant (2007), for instance, has sought a thermodynamic theory of economics, considering economic equivalents for the ideal gas equation, pressure, temperature, entropy, and the first and second laws of thermodynamics. In that work, he developed relations between utility and the law of diminishing marginal utility. He also constructed a trade cycle based on the Joule cycle, accounting for cycle efficiency, growth and entropy gain. Furthermore, Bryant describes a thermodynamic model of a money system, and thermodynamic-based measures for economic value.

Many researchers have noted that the thermodynamic quantity exergy, which stems from the second law of thermodynamics, is often a consistent measure of economic value, while energy is not. Consequently, many have recommended that accounting and pricing are better performed when based on exergy rather than energy. For instance, costs based on exergy are more rationally distributed among outputs, since exergy-based unit costs are more meaningful than energy-based ones.

Several exergy-based economic-analysis techniques have been developed. The goals of such methods usually include determining the appropriate allocation of economic resources for optimal or improved design and operation, or the economic feasibility and profitability of a system. Existing exergy-based economic techniques include exergoeconomics, thermoeconomics, exergy-based pricing, EXCEM analysis and analysis based on the ratio of thermodynamic loss to capital cost. In the latter approach, which identifies as important the ratio of thermodynamic loss rate to capital cost, devices in systems are examined and correlations are identified between capital costs and specific second-law-based thermodynamic losses (i.e., total and internal exergy losses). Such correlations can improve understanding and aid design efforts.

In this chapter, the relations between exergy and economics are discussed, and many of the analysis techniques which integrate exergy and economics are reviewed. Several illustrative examples are considered, including modern electrical generating stations operating on various fuels (coal, oil, uranium).

2. Exergy

This section describes exergy and exergy analysis, as well as their benefits. Aspects of thermodynamics relevant to exergy analysis are discussed, as are exergy principles, quantities, efficiencies and implications. A procedure for exergy analysis is provided.

2.1. The Problem with Energy Analysis

Thermodynamics permits the performance and efficiency of energy systems to be described. Conventional thermodynamic analysis is based primarily on the conservation of energy principle embodied in the first law of thermodynamics. An energy analysis of a process or system essentially accounts for the energy exiting (with products and wastes) and entering.

Efficiencies usually evaluated as energy ratios are conventionally used to evaluate and compare processes and systems.

But energy efficiencies do not always assess how nearly performance approaches ideality and are consequently often misleading. Also, factors which cause performance to deviate from ideality (i.e., thermodynamic losses) are often not properly described qualitatively and quantitatively with energy analysis. For example, energy analysis can locate the principal inefficiencies wrongly in a system and assess a state of technological efficiency different than actually exists.

2.2. Why Exergy Analysis?

Exergy analysis, a thermodynamic analysis technique based on the second law of thermodynamics, overcomes many of the shortcomings of energy analysis. Exergy analysis gives efficiencies which provide a true measure of how nearly actual performance approaches ideality, and identifies properly the causes, locations and magnitudes of inefficiencies.

Exergy analysis quantitatively indicates

- the theoretical limitations imposed on a system, which show that a real system can not conserve exergy and that only a portion of the input exergy can be recovered, and
- practical limitations by evaluating losses which directly measure lost exergy.

By providing an illuminating, rational method for meaningfully assessing and comparing systems, exergy analysis can assist in improving and optimizing designs.

Exergy analysis is described extensively elsewhere (e.g., Dincer and Rosen, 2007; Gaggioli, 1983; Moran and Shapiro, 2007; Kestin, 1980; Moran, 1989; Kotas, 1995).

2.3. What is Exergy?

The exergy of an energy or material quantity is a measure of its usefulness or quality. Technically, exergy is the maximum work obtainable from an energy or material quantity as it passes reversibly to the environmental state, exchanging heat and materials only with the surroundings (Gaggioli, 1998; Kestin, 1980; Gaggioli and Petit, 1977; Szargut, 1980).

The units of exergy are the same as those of energy, but while energy focuses on quantity, exergy focuses on quality. Since energy can be degraded in quality even though energy is conserved in quantity, exergy is not conserved for real processes.

Although energy cannot be created or destroyed, exergy (or energy quality) can be degraded, eventually reaching complete equilibrium with the surroundings and becoming of no further use for performing tasks.

For a process, the input exergy is a measure of the maximum potential associated with the input material and energy. This maximum is retained and recovered only if the inputs undergo processes reversibly. No further useful exergy can be extracted by allowing a system and its environment to interact if they are in equilibrium. Losses in the potential for exergy recovery occur in the real world because actual processes are always irreversible.

2.4. Exergy Quantities

Both material and energy quantities possess exergy. The exergy of several non-material energy quantities are considered:

- The exergy associated with shaft work, based on the definition of exergy, is equal to the energy.
- The exergy associated with electricity is equal to the energy.
- The exergy associated with a heat transfer, or the thermal exergy, is a function of the temperature at which the heat transfer occurs.

The exergy of material quantities are now described:

- The exergy of a system containing material can be expressed as the sum of the physical, chemical, kinetic, potential and other exergy terms. The exergy depends on the properties of the material and of the reference-environment, combining the system's extensive properties with the reference-environment's intensive properties.
- The exergy of a flowing stream of matter is the sum of the exergy of a system containing matter plus the exergy associated with the flow work of the stream.

2.5. The Reference Environment

Exergy is evaluated with respect to a reference environment, the intensive properties of which partly determine the exergy of a flow or system. The exergy of the reference environment is zero, as is the exergy of a flow or system in equilibrium with it.

The reference environment is in theory in stable equilibrium, with a constant intensive state (temperature pressure, chemical potentials) and all parts at rest relative to one another. The reference environment acts as an infinite system, and is a sink and source for heat and materials, but chemical reactions do not occur between the environmental components.

Table 1. A reference-environment model that emulates the natural environment

Property		Value
Temperature, T_o		25°C
Pressure, P_o		1 atm
Composition	Gaseous	Atmospheric air saturated with H_2O at T_o and P_o, with the following constituent molar make up: 75.67% nitrogen (N_2) 20.35% oxygen (O_2) 3.03% water (H_2O) 0.91% argon (Ar) 0.03% carbon dioxide (CO_2) 0.01% hydrogen (H_2)
	Solid and liquid	Condensed phases at T_o, P_o: Water (H_2O) Limestone ($CaCO_3$) Gypsum ($CaSO_4{\cdot}2H_2O$)

The natural environment does not possess the theoretical characteristics of a reference environment because: it is not in equilibrium, its intensive properties vary spatially and temporally, and many chemical reactions in the natural environment are blocked because the transport mechanisms necessary to reach equilibrium are too slow at ambient conditions.

Models for the reference environment are used which try to achieve a compromise between the theoretical requirements of the reference environment and the actual behavior of the natural environment. A common reference-environment model is the natural-environment type, which simulates the natural environment (Baehr and Schmidt, 1963; Gaggioli and Petit, 1977; Rodriguez, 1980). Normally, the temperature and pressure of this reference environment (see Table 1) are 25°C and 1 atm, respectively, and the chemical composition includes air saturated with water vapor, and several condensed phases at 25°C and 1 atm (e.g., water, gypsum, limestone). Other classes of reference-environment models have been proposed:

- In process-dependent models, only components in the process are considered, in a stable equilibrium composition at the temperature and total pressure of the natural environment (Bosnjakovic, 1963). This model is not general and does not resemble the natural environment.
- In equilibrium models, all the materials in the atmosphere, oceans and a layer of the crust of the earth are blended and an equilibrium composition calculated for a given temperature (Ahrendts, 1980). The selection of the thickness of crust considered is subjective but intended to include all materials accessible to technical processes. For thicknesses from 1 m to 1000 m at 25°C, the model does not simulate well the natural environment.
- In constrained-equilibrium models (Ahrendts, 1980), the equilibrium composition in the previous point does not permit the formation of nitric acid and its compounds, so chemical reactions that form these substances are in constrained equilibrium, while other reactions are in unconstrained equilibrium. For a crust thickness of 1 m at 25°C, the model simulates reasonably the natural environment.
- In the natural reference-substance model, a reference substance is selected (usually the least valuable substance abundantly found in the natural environment) for every chemical element and assigned zero exergy (Szargut, 1967). The selection of reference substances is primarily economic but simulates reasonably the natural environment.
- In the arbitrary reference-substance model, reference substances are selected arbitrarily and do not simulate the natural environment (Sussman, 1980, 1981).

Other characteristics of the reference-environment models and the sensitivities of exergy values to different models have been reported (Wepfer and Gaggioli, 1980; Sussman, 1981; Ahrendts, 1980).

2.6. Energy and Exergy Balances

A balance for a quantity in a system states that the accumulation of the quantity is the sum of the input and generation less the output and consumption of the quantity. Input and output refer respectively to quantities entering and exiting across system boundaries, while generation and consumption refer respectively to quantities produced and consumed in the system. Accumulation refers to build-up (either positive or negative) in the system of the quantity.

This general balance may be written for energy and exergy, as well as other quantities.

Energy, being conserved (neglecting nuclear reactions), can be neither generated nor consumed. Consequently, an energy balance can be written as

$$\text{Energy input} - \text{Energy output} = \text{Energy accumulation} \quad (1)$$

The output term in this balance can be separated into product and waste components as follows:

$$\text{Energy output} = \text{Product energy output} + \text{Waste energy output} \quad (2)$$

Exergy is consumed due to non-idealities, in proportion to the entropy creation due to irreversibilities. Therefore, a corresponding exergy balance can be expressed as

$$\text{Exergy input} - \text{Exergy output} - \text{Exergy consumption} = \text{Exergy accumulation} \quad (3)$$

As with energy, the exergy output can be divided into product and waste terms:

$$\text{Exergy output} = \text{Product exergy output} + \text{Waste exergy output} \quad (4)$$

The exergy balance is a combination of the conservation law for energy and non-conservation law for entropy. Exergy consumption can also be expressed as the difference between the total exergy flows into and out of the system, less the exergy accumulation in the system. A key difference between energy and exergy is seen with these balances: exergy, a measure of energy quality or work potential, can be consumed, but energy is conserved.

For a cyclic process with identical initial and final states, the accumulation terms in the balances are zero.

2.7. Efficiencies

Efficiency is an important factor in decisions regarding resource utilization. Many have examined efficiencies and other performance measures (Sussman, 1981).

Efficiency is defined as 'the ability to produce a desired effect without waste of, or with minimum use of, energy, time, resources, etc.,' and commonly means the effectiveness with which resources are used to do our make something, or the degree to which the ideal is approached in such activities.

For general engineering systems, efficiencies are typically ratios of appropriate quantities. For energy-intensive and energy-conversion systems, efficiencies are conventionally energy ratios. A process then has maximum efficiency if energy input equals recoverable energy output or if no 'energy losses' occur. However, such efficiencies do not measure 'approach to ideality' and so are misleading

More meaningful efficiencies require a quantity for which ratios can measure approach to ideality. The second law must be invoked as it states that ideality is attained for a reversible process, but this law must be quantified first. Although this can be done with entropy, it is difficult to create efficiency measures using entropy.

Alternatively, meaningful efficiencies are obtained with ratios based on exergy, and maximum efficiency is attained for a process which conserves exergy. Exergy efficiencies clearly measure approach to an ideal.

Gaggioli (1983) calls exergy efficiencies 'real' or 'true' efficiencies, while calling energy efficiencies approximations to real efficiencies. Some significant benefits of exergy efficiencies:

- they provide more illuminating insights into efficiency than energy efficiencies because they weigh energy flows according to their exergy contents and separate inefficiencies into effluent losses and irreversibilities,
- they provide a measure of potential for improvement,
- they are always between 0% and 100%, and are thus more intuitive than energy efficiencies, and
- they clarify measures which can be greater than 100% when energy is considered, such as coefficient of performance, as such measures are between 0% and 100% based on exergy.

There exist several exergy-based efficiencies. A common energy efficiency η for processes and systems is

$$\eta=\frac{\text{Energy in product outputs}}{\text{Energy in inputs}}=1-\frac{\text{Energy loss}}{\text{Energy in inputs}} \tag{5}$$

The corresponding exergy efficiency ψ is

$$\psi=\frac{\text{Exergy in product outputs}}{\text{Exergy in inputs}}=\frac{\text{Exergy loss plus consumption}}{\text{Exergy in inputs}} \tag{6}$$

2.8. Procedure for Energy and Exergy Analyses

A simple procedure for performing energy and exergy analyses of a process or system involves the following steps:

- Separate the process or system into as many sections as desired so as to achieve the depth of detail and understanding desired.
- Determine all basic quantities (e.g., work, heat) and properties (e.g., temperature, pressure).
- Perform mass and energy balances.

- Select an appropriate reference-environment model, accounting for the nature of the process, the acceptable accuracy, the acceptable analysis complexity, and the questions being investigated.
- Evaluate energy and exergy values, relative to the selected reference-environment model.
- Perform exergy balances, and determine exergy consumptions.
- Evaluate efficiencies, taking into account the desired measure of merit in selecting the efficiency.
- Draw appropriate interpretations, conclusions and recommendations from the results, addressing as appropriate design changes, retrofit plant modifications, etc.

2.9. Implications of Exergy Analyses

Exergy analyses of systems impact application decisions and research directions. In the latter case, exergy analyses provide insights into the 'best' directions for research, more than energy analyses, where 'best' means the most promising for significant efficiency gains. Two primary reasons for this observation follow:

- Exergy losses represent true losses of the potential to generate the desired product from the input. Energy losses do not. To increase efficiency while accounting for energy degradation, addressing exergy losses focuses research on reducing losses that impact the objective.
- Exergy efficiencies always measure how nearly a process or system approaches the theoretical upper limit. Energy efficiencies do not. By focusing research on sections or processes with low exergy efficiencies, effort is directed where the largest margins for efficiency improvement inherently exist, while focusing on energy efficiencies may target areas with little margins for improvement, even theoretically.

Application and research decisions must account for economic, environmental, safety, social, political and other factors, in addition to energy and exergy. The latter inform beneficially such decision making, and suggest from a thermodynamic view that exergy analyses typically find that improvement efforts should concentrate more on internal rather than external exergy losses, with a higher priority for the processes having larger exergy losses. Of course, research should also address processes having low exergy losses where cost-effective improvement measures are identified.

2.10. Recognition and Application of Exergy Analysis

Increasing application and recognition of the usefulness of exergy methods by those in industry, government and academia has been observed in recent years (Moran, 1989; Kotas, 1995; Edgerton, 1982; Szargut et al., 1988). Exergy methods have even been suggested as an aid for energy planning and policy development. The historical evolution of exergy analysis has been summarized by Rezac and Metghalchi (2004).

The significance of exergy has been described well by Sciubba (2001), who noted that energy conversion systems were subject to assessment based on second law concepts over five decades ago. He wrote, "relevant design procedures of the time neglected to recognize that the irreversibility in processes and components depend on the energy 'degradation rate' and not only on the ratio between the intensities of the output and input flows, and that there is a scale of energy quality that can be quantified by an entropy analysis." The legacy of this early work, which led to the development of exergy analysis, is that efficiencies based solely on energy and the first law are increasingly recognised as at best misleading and often erroneous. Sciubba further notes that exergy analysis "has had a very profound impact on the energy conversion system community, to a point that it is difficult today to find a design standard which does not make direct or indirect use of exergetic concepts in its search for an 'optimal' configuration."

Numerous applications of exergy analysis to improve efficiencies of processes and systems have been reported, ranging in size from simple to complex processes. Some examples follow:

- electricity generation using conventional technologies like fossil fuel and nuclear power plants (Rosen, 2001) as well as advanced technologies like fuel cells (Cownden et al., 2001), pressurized fluidized bed combustion (Rosen and Horazak, 1995) and geothermal plants (Dipippo, 1994),
- cogeneration of electricity and heat (Rosen et al., 2005),
- thermal processes (Bejan, 1982; Kotas, 1995; Szargut et al., 1988), including drying (Dincer, 2002), industrial heating (Rosen and Dincer, 2004), district heating (Ozgener et al., 2004), cooling (Kanoglu et al., 2004), dehumidification (Van Den Bulck et al., 1988), fluid flow (Bejan, 1982) and thermal energy storage (Rosen and Dincer, 2002; Rosen et al., 2004),
- chemical, and metallurgical processes (Szargut et al., 1988; Sato, 2005),
- renewable energy utilization (Koroneos et al., 2003), including geothermal (Lee, 2001; Dipippo, 1994),
- industrial processes (Brodyanski et al., 1994), and
- transportation, including aerospace (Rosen and Etele, 2004) and automotive (Cownden et al., 2001).

The incorporation of exergy into engineering design has also received increasing attention (Bejan et al., 1996; Gaggioli, 1983).

Exergy methods have been applied to larger systems like countries and sectors of their economies. For example, analyses have been carried out for regions and countries in:

- North America, including Canada (Rosen, 1992) and the United States (Reistad, 1975; Ayres et al., 2003),
- Europe, including the United Kingdom (Hammond and Stapleton, 2001), Sweden (Wall, 1997), Italy (Wall et al., 1994) and Turkey (Ozdogan and Arikol, 1995; Rosen and Dincer, 1997b),
- the Middle East, including Saudi Arabia (Dincer et al., 2003), and
- Asia, including Japan (Wall, 1990).

As a consequence of the growing acceptance of exergy, specialized textbooks on exergy and related methods have been written (Bejan, 1982. Kotas, 1995; Moran, 1989; Szargut, 2005; Szargut et al., 1988; Brodyanski et al., 1994; Sato, 2005) and general thermodynamics texts have increasingly included material on exergy (Bejan, 2006; Moran and Shapiro, 2007).

3. Exergy and Economics

Exergy is a useful concept in economics (Tsatsaronis and Valero, 1989). Wall (2003), for instance, states that exergy is crucial not only to efficiency studies but also to cost accounting and economic analyses, and goes on to note that costs should reflect value, and since value is not generally associated with energy but with exergy, assignments of cost to energy lead to misappropriations, which are common and often gross. It has been further pointed out that exergy possesses an intrinsic and direct correlation with economic values (Sciubba, 2001).

This duality of exergy is important. On the one hand, exergy allows the rational evaluation of the value of fuels and resources, as well as the efficiencies of processes and devices and their inefficiencies (including internal exergy consumptions due to dissipation or irreversibility and external waste exergy emissions). On the other hand, exergy facilitates the determination of the costs associated with thermodynamic losses and the values and costs of the outputs and accumulated quantities of processes and devices.

Consequently, using exergy content as a basis for cost accounting can help in determining the real costs of producing commodities and in pricing such products. In addition, exergy can help evaluate economic viability and profitability. The benefits and drawbacks of exergy-based economic methods, as well as the relations between exergy and economics, have been investigated (Rosen, 2002b).

Exergy can interface broadly with economics. In microeconomics, exergy can be combined with cost-benefit analyses to improve designs. In macroeconomics, exergy provides a basis for increasing efficiency, or reducing resource utilization and losses, or reducing environmental damage. These objectives can be accomplished with, for example, exergy-based incentives or exergy taxes.

Since exergy can be utilized in concert with macroeconomics and microeconomics to provide useful methods for the evaluation and improvement of systems, exergy more broadly can help in optimizing engineering designs and in making operating decisions. Wall (2003) describes exergy-based concepts and methods for analyzing and optimizing energy systems, including exergy-economic optimization as well as exergy utility diagrams and life cycle exergy analysis.

Tsatsaronis and Park (2002) have examined the relation between exergy destructions and investment costs in thermal systems.

Jaber et al. (2004) have attempted to link thermodynamic concepts, including exergy and entropy, to price-driven economic systems. Related investigations involve modelling the economic order quantity with an approach based on the second law of thermodynamics (Jaber et al., 2006) and using entropy cost to evaluate the economic order quantity for repair and waste disposal (Jaber and Rosen, 2008). Also, Nuwayhid et al. (2006) consider diffusion-like economic commodity flows from a thermodynamic perspective, by modelling a commodity flow as a quasi-thermodynamic irreversible process. In that work, which supposes that the laws of thermodynamics can explain the process of commodity diffusion, a coupling relationship between commodity price and quality is derived.

In the remainder of this section, cost balances and cost allocation are described and illustrated, and the inclusion of environmental costs is considered.

3.1. Cost Balance

Like for thermodynamic quantities, a balance can be written for cost:

$$\text{Cost input} + \text{Cost generation} - \text{Cost output} = \text{Cost accumulation} \tag{7}$$

Cost input and output represent respectively the cost associated with all inputs and outputs for the system. Cost accumulation represents build-up or accumulation of cost for the system, while cost generation corresponds to the appropriate capital and other costs associated with the creation and maintenance of a system.

The cost generation term highlights the fact that cost is a non-conserved quantity. Further, cost generation is always positive since cost is an increasing quantity. Cost generation can be written as follows:

$$\text{Cost generation} = \text{Capital cost of equipment} + \text{Other creation and maintenance costs} \tag{8}$$

where the capital cost of equipment K is the portion of cost generation associated with the cost of equipment. Other creation and maintenance costs include interest and insurance costs.

Cost generation is sometimes approximated by the capital cost of equipment. With this simplification, the costs associated with inputs and the cost allocations among outputs are not required. This simplification is sometimes justifiable because

- capital cost is often the most significant component of cost generation, so the consideration of only capital cost closely approximates cost generation, and
- cost generation components other than capital costs are often proportional to capital cost. Hence the simplification can be reasonable when qualitative trends are sought.

The simplification is usually employed when use of the cost generation term increases excessively the complexity of analysis, since numerous other economic details are required like lifetimes, salvage values, interest rates, etc.

The above cost balance can be applied on rate and quantity bases. In a rate-based cost balance, the cost generation rate represents the total cost generation levelized over the operating life of the system. In a quantity-based cost balance, the cost creation quantity represents the portion of the total cost generation accounted for during the time period considered.

3.2. Comparison of Economic Cost and Thermodynamic Balances

Although economic cost and thermodynamic balances appear similar, they exhibit several significant differences:

- Energy is conserved but economic cost is not.
- Both cost and exergy are not conserved, with cost generated due to the cost of devices and exergy consumed due to irreversibilities associated with their operation.
- In energy and exergy balances, the values associated with all quantities are defined by thermodynamics. Only cost input and generation are defined for the cost balance, however; the distribution of costs over outputs and accumulations can be made subjectively. This point is elaborated upon in the following two subsections.

3.3. Cost Allocation

The fact that costs may be distributed among outputs and accumulations in multiple ways is important and often the cause of confusion, particularly when there is more than one product. The allocation of costs often depends on such factors as the type and purpose of device and other economic considerations. If costs are not allocated appropriately, it is difficult to gauge the cost-effectiveness of a device and to set prices for salable commodities.

Confusion arises most commonly for devices for which there are multiple products. In cogeneration systems which produce multiple products, costs can be allocated entirely to one of the products, or divided among the products. If this determination is not made properly, products can be priced inappropriately and the profitability of a device or process may be incorrectly assessed.

Sometimes a pseudo-scientific approach to cost allocation is taken and costs are allocated proportionally to all outputs and accumulations of a quantity (such as mass, energy or exergy). Alternatively, costs are sometimes allocated proportionally to non-waste outputs and accumulations of a quantity. Although simplistic, such approaches often do not allocate costs in ways considered to be realistic and reasonable relative to market forces and industry perceptions.

3.4. Illustrations

Consider as an example two plants: a thermal power generation plant and a cogeneration, or combined heat and power (CHP), plant.

For the thermal power plant, we assume there is no storage of outputs within the plant. Subjective decisions must be made regarding the allocation of costs among outputs, which can include the electrical product, by-products and wastes. Waste heat may simply be treated as a waste, in which case it is reasonable to allocate none of the costs to it. The only product is electricity, so it makes sense to allocate all of the costs to it.

For the cogeneration plant, the allocations depend on the uses for the outputs. If the heat is not a waste because its temperature is sufficiently high to be desirable to heat users, it may be reasonable to allocate part of the total costs to it, and the remainder to the product electricity. Then, the heat is in essence treated as a by-product or a co-product.

Returning to the thermal power, it is noted that it can be modified to become a cogeneration plant. This modification usually entails sacrificing some electrical power to raise the temperature of the waste heat so that it becomes purchasable. In this case, the allocation of costs would change from what was appropriate when the plant only produced electricity.

3.5. Extending the Links Between Exergy and Economics to Environmental Aspects

Exergy has been found to be useful in understanding and assessing environmental impact (Sciubba, 1999; Tribus and McIrivne, 1971; Rosen and Dincer, 1997a, 1999b; Gunnewiek and Rosen, 1998; Rosen, 2002a), non-renewable resource depletion (Szargut et al., 2002), ecology (Szargut et al., 2002; Szargut, 2005; Jorgensen, 2000; Jorgensen and Fath, 2004) and sustainable development (Dincer and Rosen, 2007). Exergy has also been found to relate to environmental methods like industrial ecology (Connelly and Koshland, 2001a, 2001b; Dewulf and Van Langenhove, 2002; Kay, 2002).

With the relations described in these references, the ties between exergy and economics can be extended to environmental aspects. This can be accomplished in several ways:

- The characteristic of exergy that allows it to be an indicator of potential environmental impact can be incorporated in exergy-based economic assessments.
- The costs associated with environmental impact and protection can be included in the economic portion of an exergy-based economic assessment.

By extending exergy and economics to account for environmental effects, techniques can developed that permit one to minimize life cycle costs and thereby to find the most appropriate device or process. This can be accomplished while reducing environmental effects, for given prevailing economic conditions.

4. Exergy and Economic Methods

A brief summary is presented here of existing analysis techniques which integrate exergy and economics, and which have been developed by researchers over the last several decades. These methods for economic analyses based on exergy include thermoeconomics, second-law costing, cost accounting, exergoeconomics and others. Many researchers of the thermodynamic and economic aspects of energy systems cite Georgescu-Roegen (1971) as a pioneer in the field of the thermodynamics of economics and the father of the discipline.

Most exergy-based economic analysis methods have several common characteristics:

- they combine exergy and economics to help achieve thermodynamic and economic objectives, and
- they recognize that exergy, not energy, is the commodity of value in a system, and they consequently assign costs and/or prices to exergy-related variables.

The goals of most exergy-based economic techniques include one or more of the following:

- assessment of the economic feasibility and profitability of a system, in part by determining the actual costs of products and appropriate prices; and
- evaluation of the appropriate allocation of economic resources so as to optimize the design and operation of a system.

The use of thermoeconomics for system improvement has been investigated by many (El-Sayed and Tribus, 1983). Further, the optimization aspect embedded within the latter point is particularly important in the application of exergy-based economic techniques (Hua et al., 1989). Optimal designs need to meet multiple requirements, including performing as specified by design data and satisfying all constraints, and displaying the most advantageous characteristics under a given set of conditions. Noting that optimization can be challenging, Sciubba (2001) points out optima are not always expressed by a well-posed mathematical objective function and instead often involve vague or incomplete optimization design criteria which are often fuzzy.

Four main categories of exergy-based economic methodologies are identified, depending on which of the following forms the basis of the technique (Tsatsaronis, 1987): 1) exergy-economic cost accounting, 2) exergy-economic calculus analysis, 3) exergy-economic similarity number, and 4) product/cost efficiency diagrams. Reistad and Gaggioli (1980) have developed methods for available energy costing. Further information on methods of economic analyses based on exergy have been reported in textbooks (e.g., Kotas, 1995; Szargut et al., 1988; Bejan, 1982), in critical reviews and comparisons (e.g., Tsatsaronis, 1987; Kotas, 1995; Bejan, 1982; Rosen and Dincer, 2003), and in numerous research articles.

4.1. Thermoeconomics and Exergoeconomics

Exergoeconomics and thermoeconomics are relatively well known exergy-based economic-analysis methodologies that have evolved over several decades (Yantovskii, 1994; El-Sayed, 2004). Exergy and microeconomics forms the basis of thermoeconomics, which is also called exergoeconomics (Bejan et al., 1996) and exergonomics (Yantovskii, 1994). The concept of utility is a central concept in macroeconomics and is also closely related to exergy. An exergy tax is an example of how exergy can be introduced into macroeconomics.

Thermoeconomics is based on a combination of economic and thermodynamic methods, and has been developed to meet industry requirements and standards. With this tool, efficiencies are determined with exergy analysis, while non-energy expenditures such as financial and labour costs are related to the technical parameters of the device under consideration. Thermoeconomics can be used in optimisation activities (Gogus, 2005). In thermoeconomic optimization, the optimal design point and operating conditions are determined by minimising the total economic cost under financial, technical, environmental and other constraints. Sciubba (2001) states that thermoeconomics provides "a complete and theoretically sound tool."

Tsatsaronis (1996) describes thermoeconomic analysis and its use in the optimization of energy systems. A systematic and general methodology for evaluating costs in thermal systems is described by Lazzaretto and Tsatsaronis (2006).

Algorithms for multi-objective energetic and economic optimization in thermal systems design are described by Toffolo and Lazzaretto (2002). The algorithms allow multi-objective optimization that finds solutions that simultaneously satisfy exergetic and economic objectives. The approach is important because considering only economic minima may not be sufficient for sound decision making; more efficient solutions may result in more advantageous designs due to changes in energy market prices or in energy policies, in spite of small increases in total costs.

Exergy accounting utilizes exergetic costs and can be used for diagnosing energy systems and for accounting for natural exergy resources (Verda et al., 2005a, 2005b). The method measures the amount of exergy resources to produce a good. Costs must be properly formed to understand and evaluate exergy costs and the degradation of resources entering a system, as well as the relations between cost and irreversibility. Capabilities and drawbacks of the field of exergy cost accounting have been explained by Valero (2006).

Valero et al. (2006a) describe the theoretical fundamentals and present numerous applications of exergy cost accounting and thermoeconomics. The basic requirements are explained to calculate exergy and thermoeconomic costs and to perform thermoeconomic analyses of complex systems. The authors also explain how the approach can be used for operation diagnosis and optimization of a complex system.

Erlach et al. (1999) examine the use of structural theory in thermoeconomics, and Torres et al. (2002) discuss malfunction and dysfunction analysis using structural theory within thermoeconomics.

Mazur (2005) includes uncertainty in thermoeconomic analysis in order to find solutions that simultaneously satisfy thermodynamic and economic objectives. That work considers fuzzy nonlinear programming where local criteria like maximum efficiency and minimum total cost rate as well as different constraints in an ill-structured situation can be represented by fuzzy sets. Examples of fuzzy thermoeconomic analysis are presented by the author.

4.2. EXCEM Analysis

EXCEM (exergy, cost, energy and mass) analysis is a methodology that focuses on the four key parameters represented by its name for evaluating devices and processes. Developed by the author (Rosen and Dincer, 2003c), EXCEM analysis combines exergy and economics and can be useful for the assessment and improvement of systems and processes.

4.2.1. EXCEM Framework and Rationale

The rationale of EXCEM analysis is relatively straightforward: an understanding of the performance of a system or process necessitates that the flows of the quantities exergy, cost, energy and mass into, out of and at all points within a system be examined.

The framework for EXCEM analysis is illustrated generally in Figure 1a and for an example in Figure 1b. Balances can be written for each of the EXCEM quantities. Of the quantities represented by EXCEM, exergy and cost are not conserved. Exergy decreases or remains constant, while cost increases or remains constant. Only mass and energy are conserved.

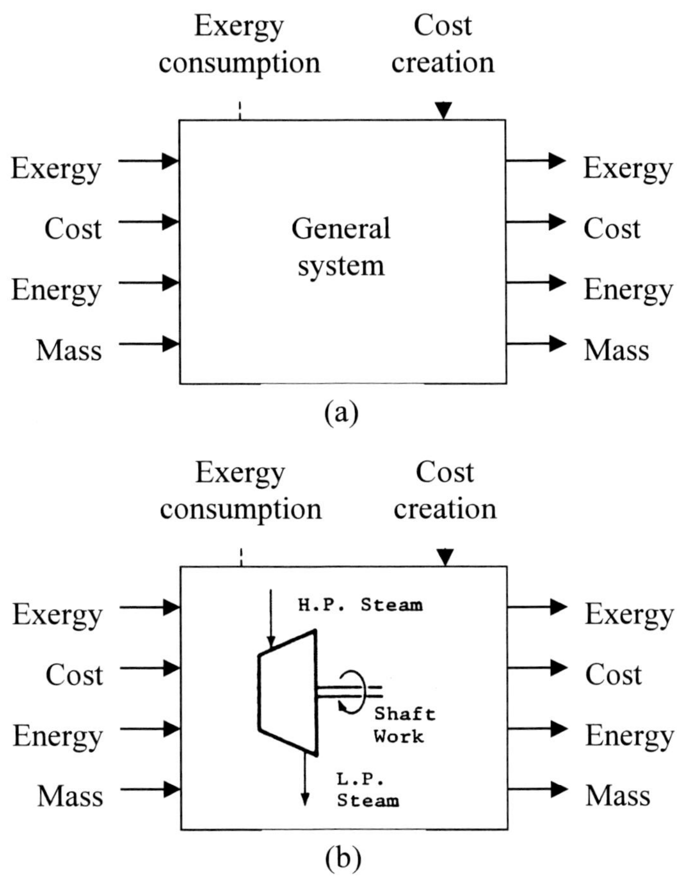

Figure 1. Illustration of EXCEM analysis structure and its application for (a) a general system and (b) a steam turbine. Exergy consumption and cost creation are shown with dashed lines to denote that they do not actually enter or exit the system boundary.

EXCEM analysis was developed to form the basis of a unified methodology for thermodynamic, economic and environmental decisions, and to assist in design.

EXCEM analyses of numerous engineering processes have been carried out, demonstrating that the EXCEM analysis methodology can provide valuable insights into performance and efficiency, economics, and potentials for environmental damage for systems and processes. Interestingly, the exergy-related aspects of EXCEM are often the most informative.

Previous analyses have also demonstrated the findings of EXCEM analyses are sometimes not obtainable through conventional energy and economic analyses, while many findings of conventional analyses are obtained more directly with EXCEM analysis.

EXCEM analysis can be incorporated into computer codes like process simulators, making the methodology more useful and straightforward to apply. In such a form, EXCEM analysis provides accurate results in a convenient manner for a wide range of processes. One such effort by the author involved enhancing the Aspen Plus process simulator for EXCEM analysis (Dincer and Rosen, 2007). The enhancement of Aspen Plus for EXCEM analysis is general and the enhancement approach can be used on other process-simulation and related codes.

4.2.2. EXCEM Illustrations

Illustrations are presented to clarify the general concepts associated with EXCEM analysis. The focus is on cost flows, so mass, energy and exergy balances are not extensively discussed. Three devices operating at steady-state are considered:

- *Electrical transformer.* The application of EXCEM analysis to an electrical transformer, in which electricity is converted from one voltage to another, is relatively straightforward. Electricity is input and electricity at a different voltage and waste heat are output. In this example, the principal input and output flows are the same energy form (electricity), so energy and exergy analyses are very similar. If the transformer is adiabatic (closed to heat flows) all cost associated with inputs and generation is allocated to the one output. If the transformer is not adiabatic, all input and generation costs are still logically allocated to the output electricity, because the heat loss is a waste.
- *Electrical generator.* In an electrical generator, mechanical work in the form of a rotating shaft is converted to electricity. Waste heat is also output. Here, the main input and output flows (work and electricity) are theoretically equivalent in quality, so energy and exergy analyses are very similar. EXCEM analysis can therefore be straightforwardly applied to the electrical generator. Whether or not the electrical generator is adiabatic, it is sensible to allocate all input and generation costs to the output electricity and none to the waste heat.
- *Steam turbine.* A steam turbine generates work in the form of a rotating shaft from an input flow of high-pressure steam. Exhaust steam exits from the low-pressure side of the turbine, and extraction flows of steam sometimes exit the turbine at intermediate pressures to meet thermal demands. The application of EXCEM analysis to a steam turbine (Figure 1b) is more complex than for the two previous cases because a subjective decision must be made regarding the allocation of costs. The shaft work, exhaust steam and extraction steam flows may each be allocated part of the input and generation costs. The allocations depend on the uses for the outputs. The exhaust steam, for example, may or may not be a waste.

4.2.3. Exergy and Cost Relations

The relation between exergy and cost can be further illustrated by plotting exergy loss as a function of cost generation. Both internal exergy losses (i.e., consumptions) or total exergy losses (i.e., consumptions plus waste emissions) can be considered.

Such plots demonstrate the relation between cost generation and exergy losses, highlighting the fact that financial investments must usually be made to avoid losses which diminish system performance from ideality.

Several notes are worth making on the generation of such plots. First, when total exergy losses are considered, the intensive properties of the reference environment must be completely specified. Second, when only internal exergy losses are considered, the reference environment need not be specified completely as only its temperature is required. Third, costs associated with inputs need not be specified.

4.3. Loss-Cost Ratio Analysis

Loss-cost ratio analysis is an exergy-based economic method that focuses on the ratio of thermodynamic loss rate to capital cost. The insights provided with loss-cost ratio analysis can assist in integrating thermodynamics and economics so as to assist in analysing and designing energy systems (Rosen, 1990, 1991).

Loss-cost ratio analysis involves using data for devices in systems to determine such ratios and to identify correlations between capital costs and specific second-law-based thermodynamic losses (i.e., total and internal exergy losses).

The approach has been applied previously to mature devices. Examinations of thermodynamic and economic data identified correlations between capital costs and thermodynamic losses for devices. The relations between energy losses and capital costs generally were observed not to provide helpful information.

The existence of correlations between capital costs and exergy-based thermodynamic losses for systems and their component devices implies that designers incorporate the recommendations of exergy-based economic methods into process designs, either directly and overtly, or indirectly and unknowingly (Rosen, 1990, 1991, 2002b; Rosen and Dincer, 2003a).

In loss-cost ratio analysis, balances for appropriate quantities are considered, following Eqs. 1 and 3.

4.3.1. Key Loss Terms

Several types of thermodynamic losses are considered in loss-cost ratio analysis. These are defined here, using rate forms of the thermodynamic balances described previously (see Eqs. 1 and 3).

To simplify the present discussions, only devices are considered where no losses are associated with the accumulation terms in the energy and exergy balances.

Exergy losses can be identified from exergy balances (see Eqs. 1 and 3). There are two types of exergy losses, which sum to the total exergy loss. The "waste exergy output" represents the loss associated with exergy emitted from the system, and the "exergy consumption" represents the internal exergy loss due to process irreversibilities. Loss rates can be written for each of these exergy losses. That is, the loss rate based on internal exergy consumption, $L_{ex\text{-}i}$, is defined as

$$L_{ex\text{-}i} \equiv \text{Exergy consumption rate} \tag{9}$$

while the loss rate based on external exergy losses or waste emissions, $L_{ex\text{-}e}$, is defined as

$$L_{\text{ex-e}} \equiv \text{Waste exergy output rate} \tag{10}$$

and the loss rate based on total exergy loss, L_{ex}, is defined as

$$L_{\text{ex}} \equiv L_{\text{ex-i}} + L_{\text{ex-e}} \tag{11}$$

For comparison, energy losses are also sometimes considered. Energy losses can be identified directly from an energy balance, where all losses are seen to be associated with waste outputs since energy is conserved. For a system, therefore, the energy loss rate, or the loss rate based on energy L_{en}, is simply the "waste energy output rate." That is,

$$L_{\text{en}} \equiv \text{Waste energy output rate} \tag{12}$$

For a system operating in a steady-state mode, the accumulation terms in the energy and exergy balances are zero. Hence the energy and exergy loss rates described above can be determined as follows:

$$L_{\text{ex-i}} = \sum_{\text{devices}} \text{Exergy consumption rates} \tag{13}$$

$$\text{L}_{\text{ex-e}} = \sum_{\text{wastes}} \text{Exergy flow rates} \tag{14}$$

The loss rate based on exergy can be evaluated with Eq. 11 or as follows:

$$\text{L}_{\text{ex}} = \sum_{\text{inputs}} \text{Exergy flow rates-} \sum_{\text{products}} \text{Exergy flow rates} \tag{15}$$

The loss rate based on energy can be determined as

$$\text{L}_{\text{en}} = \sum_{\text{inputs}} \text{Energy flow rates-} \sum_{\text{products}} \text{Energy flow rates} \tag{16}$$

where the summations in the previous two equations are over all input streams and all product output streams.

4.3.2. Capital Costs

Values of capital cost K for the equipment in a system are considered here. Such other costs as those associated with interest and insurance are not neglected for simplicity.

4.3.3. Ratio of Thermodynamic Loss Rate to Capital Cost

Using the capital cost K from earlier and the above types of thermodynamic loss rate L, the ratio R can be defined as

$$R \equiv L/K \tag{17}$$

The value of R generally depends on whether it is based on internal exergy loss rate $R_{ex\text{-}i}$, external exergy loss rate $R_{ex\text{-}e}$, total exergy loss rate R_{ex}, or energy loss rate R_{en}. These terms are defined as follows:

$$R_{ex\text{-}i} \equiv L_{ex\text{-}i}/K \tag{18}$$

$$R_{ex\text{-}e} \equiv L_{ex\text{-}e}/K \tag{19}$$

$$R_{ex} \equiv L_{ex}/K \tag{20}$$

$$R_{en} \equiv L_{en}/K \tag{21}$$

Although values of the parameter R based on total, internal and external exergy loss rates and on energy loss rate are often considered in loss-cost ratio analysis, the ratios based on total and internal exergy loss rates are normally the most useful.

A systematic correlation often exists between exergy loss rate (total or internal) and capital cost for processes and systems. Furthermore, a correlation appears to exist between the mean thermodynamic-loss-rate-to-capital-cost ratios for all of the devices in a system and the ratios for the overall system, when the ratio is based on total or internal exergy losses, but not when it is based on energy losses. This correlation implies that devices in successful systems are configured so as to achieve an overall optimal design, by appropriately balancing the exergy-based thermodynamic and economic characteristics of the overall system and its devices.

4.4. Exergy-Based Pricing and Cost Allocation

The selection of energy sources for residential, commercial, transportation, industrial and other uses is primarily made based on prices. Consequently, it is important to determine prices appropriately using sound methods.

Incorrect pricing can lead to numerous problems. When prices are set too high, for instance, businesses can be uncompetitive with other players in the marketplace. But setting prices too low can lead to non-viable businesses that fail to cover costs. Energy systems can be uneconomical over the long-term if prices are inappropriate, as planning can then be rendered inadequate. This is often observed when prices are set based on short-sighted analyses or political assessments or on inadequate information about energy resources and the consequences of their usage.

4.4.1. Pricing and Cost Allocation Using Exergy

Prices based on exergy can be designed so as to foster resource saving and efficient technology. Prices for a commodity based on exergy usually parallel the physical value of the commodity, while prices based on energy do not exhibit this feature. Consequently, the prices of physical resources can be more rationally set based on exergy.

The use of exergy in economic valuation and the correlation of exergy with money have been considered by Bandura and Brodiansky (2001). An exergy-based determination of production expenses is shown to provide a basis for natural price determination, and a corresponding macroeconomic dynamics approach is proposed which is relatively general. The exergy model of macroeconomic dynamics was tested using US data over 25 years.

Wall (1997) notes that energy and exergy values, and the differences between them, can affect price setting for energy commodities. Wall also examined the manner in which the energy and exergy values for fuels can affect pricing.

4.4.2. Pricing of Energy Commodities Using Exergy

The varying characteristics of different fuels and energy forms are clearly seen by considering several common energy categories:

- *Work and electricity*: Electricity and work are in theory totally convertible to each other. Since work is the basis of the definition of exergy, the energy price is also the exergy price. The price of electricity varies with such factors as power capacity, as well as energy consumption and fixed and variable costs associated with its production and transmission.
- *Hydrocarbon fuels*: Hydrocarbon fuels include solids like coal and wood, liquids like oil and gasoline and gases like natural gas. For hydrocarbon fuels, the exergy and energy contents are approximately the same (typically varying by no more than 18% and usually less than 10%).
- *Heat and cold*: The exergy of a transfer of heat or cold can be evaluated as

$$Ex = Q(1 - T_0/T) \tag{22}$$

where T denotes the temperature at which the heat or cold is transferred and T_0 the reference-environment temperature. Cold transfer is simply the transfer of heat in the opposite direction. The exergy content thus varies with the reference-environment temperature, which is often modeled as the outdoor temperature. The energy and exergy cost of heat or cold are usually very different. The exergy cost depends significantly on the temperature, which determines the quality or usefulness of the thermal energy.

In some instances, the energy-conversion technology used significantly affects price. For example, one can produce heat and electricity in two independent processes, or simultaneously in a cogeneration process for heat and electricity. The determination for cogeneration of what proportion of the cost should be allocated to each of the thermal and electrical products is important to the commercial success of the process, yet there is no simple and widely accepted method for such cost allocations. If costs are not allocated appropriately, and prices set correspondingly, one product can end up overpriced (and thus not likely to be purchased) while the other product can end up under-priced (and thus likely in great demand but not be able to recover costs).

4.5. Exergy and Environmental Economics

Designing efficient and cost-effective systems, which meet environmental requirements, is one of the foremost challenges facing engineers. Given the world's finite natural resources and large energy demands, it is important to understand mechanisms which degrade energy and resources and to develop systematic approaches for improving systems while simultaneously reducing environmental impact.

The principles of mass and energy conservation indicate that raw material inputs to economic processes are not consumed, but instead ultimately return to the environment, from which they were extracted, as wastes. If the Earth is treated as a closed system, the concepts of exergy and entropy yield different economic implications, suggesting that constraints are imposed on economic growth because economic processes utilize high-exergy (or low-entropy) raw materials such as fuels and high-grade minerals, and discard low-exergy (or high-entropy) wastes. Since the Earth is an open system that receives large quantities of high-exergy (or low-entropy) solar radiation, energy resources may be adequate to sustain economic activity in the solar system over time, although non-renewable natural resources like metal ores and fossil fuels may eventually be exhausted.

Several researchers have investigated exergy and environmental economics (Edgerton, 1982).

Ayres (1998) links economics and the second law in what he refers to as "eco-thermodynamics." He argues that the economic significance of the second law lies in the fact that exergy is not conserved and is a useful measure of resource quality and quantity that is applicable to energy and materials. Ayres states that exergy is a factor of production like labor and capital, and that exergy has strong implications on economic growth theory, especially in assessments of the role of technical progress.

Sciubba (2001) extends exergy accounting and thermoeconomics with environmental factors to improve the analysis and design of energy processes and systems. The approach allows direct quantitative comparisons of factors like labour and environmental impact and externalities. Sciubba argues that the approach 1) permits some issues that are difficult to address with a purely monetary theory of value to be resolved without introducing arbitrary assumptions, and 2) is a natural development of the economic theory of commodity production, which it extends by accounting for the unavoidable exergy consumption (or energy dissipation) in the production chain. Extended Exergy Accounting exploits the correlation between exergy and economic values to develop a theory of value based on exergy or monetary metrics. The premise of the approach is that, while exergy and monetary costs may have the same morphology, in that both represent the quantity of resources that must be consumed to produce a given output, their structures may differ, allowing for different optimal designs. The approach recognizes that a theory of cost should be formulated based on exergy destruction because it conveys the notion that resources have to be consumed to produce a product or service.

Sciubba (2003, 2004) also proposes Extended Exergy Accounting as a cost analysis method for energy systems using a resource-based quantifier. The method allows the costing of production chains, and utilizes the Leontiev-type input-output technique adopted in most costing theories, including thermoeconomics. Rather than monetary units, a resource-based quantifier (extended exergy), is employed. This quantifier allows labour and financial costs to be linked to an equivalent resource consumption, thereby expressing the total exergy consumption required to produce one person-hour of work or one monetary unit of currency circulation. Environmental remediation costs are taken into account by determining the equivalent cumulative exergy expenditure required to achieve zero impact. Extended Exergy Accounting thus evaluates the resource-based value of a commodity, which is not necessarily its monetary cost, thus enabling energy planners to perform more comprehensive and meaningful assessments. The technique complements such other tools as life-cycle assessment and environmental footprint analysis and incorporates some elements of those and other environmental methods like cumulative exergy analysis and emergy analysis.

Lazzaretto and Toffolo (2004) show how energy-system designs can be optimized using separate objectives relating to energy, economics and the environment. This multi-criterion optimization is argued by the authors to be advantageous to single-objective thermoeconomic optimization and a two-objective energy and economic optimization. An environmental impact objective function is expressed in cost terms by weighting carbon dioxide and nitrogen oxide emissions according to their unit damage costs.

Tonon et al. (2006) propose a comprehensive analysis method based on energy, exergy, economic and environmental factors. Environmental emissions are considered and performance indicators developed, providing useful information about system performance and possible areas of improvement.

A method for performance evaluation under maximum ecological and maximum economic conditions is described by Tyagi et al. (2007). In this method, the ecological function is represented by the power output divided by the entropy generation rate, and the economic function by the power output divided by the total cost.

A thermoeconomic method to increase the efficient use of exergy resources based on a carbon exergy tax is proposed by Santarelli (2004), with the intent of promoting the efficient use of exergy resources. The author feels that the method, which evaluates the cost of exergy destroyed and rejected of an energy system, and which is connected with the CO_2 emissions, can improve the economic suitability of systems by permitting them to make better use of exergy. Sciubba (2005) has also proposed exergoeconomics as a thermodynamic foundation for rational resource use.

Some authors shift the focus of exergy-based economic methods that encompass environmental factors to sustainable development. For instance, Ferrari et al. (2001) discuss the integration of thermodynamics and economics so as to obtain exergy-based indicators of sustainable development.

Exergy and entropy are applied to the steady-state economy by Honkasalo (1998), so as to develop a model for sustainable development at the macro-economic level. The approach combines resource depletion with pollution and focuses environmental protection on reducing degradation losses. As a consequence, industries are able to seek alternative possibilities for environmentally sound production processes and products. The method can be applied as a conceptual tool that permits companies to set environmental goals and establish corresponding environmental programs.

5. Applications of Exergy-Based Economic Methods

Applications of methods integrating exergy and economics to many devices, systems and processes have been reported in a wide range of fields, including power generation, cogeneration, air conditioning, heating, chemical processing, separation and fuels production. In these reports, systems using conventional and renewable energy resources have been considered. Several examples of applications reported in the literature are discussed here. Applications to various systems of exergy cost accounting and thermoeconomics are presented by some researchers (Valero et al., 2006b), while others focus on individual technologies.

5.1. Power Generation

Numerous exergy-based economic investigations have been undertaken relating to electrical power generation, considering conventional and alternative power plants, hydroelectric stations, fuel cells and thermoelectric plants.

Zhang et al. (2006) applied thermoeconomics to a coal-fired power plant, and subsequently extended that work by applying thermoeconomics to the diagnosis of a coal-fired power plant (Zhang et al., 2007). Rosen and Dincer (2003a) have also used thermoeconomic analysis to investigate a coal-fired plant.

Exergoeconomic analysis has also been applied to several power plants operating on various fuels, including uranium, oil and coal (Rosen and Dincer, 2003b). Valero et al. (2002) applied thermoeconomics using structural theory with malfunction and dysfunction analysis to an actual power plant.

A complex Brayton cycle for power generation was investigated considering ecological and economic conditions by Tyagi et al. (2007), who defined the economic function as the ratio of power output to total cost and the ecological function as the ratio of power output to entropy generation rate. The cycle was optimized by adjusting several operating conditions, including cycle temperatures as well as reheat and intercooling pressure ratios. Values were determined of turbine outlet temperature and several pressure ratios at which the cycle is maximized in terms of the economic and ecological objectives while minimizing entropy generation rate.

An exergoeconomic optimisation of a Kalina power-generation cycle was carried out recently by Borgert and Velasquez (2004). In that study, the economic advantage was examined of using an ammonia-water Kalina absorption power cycle as a bottoming cycle for a cogeneration plant. The cycle was optimised using as an objective function the exergoeconomic unit cost of the electricity produced by both the cogeneration plant and the absorption cycle. Net power and unit electricity cost each decreased by 3% for the optimized system.

An exergoeconomic analysis of a 10-kW PEM fuel cell was undertaken by Kazim (2005), considering various operating conditions, e.g., temperatures, pressures, cell voltages and air stoichiometries. The effects of varying these parameters within fixed ranges on the exergy cost of the fuel cell were determined. Typical fuel cell operating voltages of 0.5 and 0.6 V were examined. The effects on the exergy cost of the fuel cell of operating pressure, cell voltage and air stoichiometry were demonstrated. A significant reduction in exergy cost was shown to be possible with lower fuel cell capital costs, annual operation and maintenance costs and hydrogen costs.

A combined power plant consisting of a solid oxide fuel cell and gas turbine was assessed using a thermoeconomic method based on a carbon exergy tax that was directed at increasing the efficient use of exergy resources (Santarelli, 2004). Also, hydroelectric and thermoelectric power generation processes were analyzed by Tonon et al. (2006) with a comprehensive method based on exergetic, economic and other parameters. Environmental emissions were considered, and performance indicators developed.

Santarelli and Macagno (2004) have undertaken a thermoeconomic analysis of a photovoltaic-hydrogen system for residential buildings.

5.2. Cogeneration and Multi-Product Generation

Several exergy-based economic investigations have been undertaken of processes that simultaneously produce multiple products, including the cogeneration of electricity and heat.

Sugimoto et al. (2006) performed exergy cost minimisation on a simple gas turbine-based cogeneration system. The optimum turbine inlet temperature for exergy cost minimisation was 1200°C. The cost of the gas turbine was observed to increase greatly at turbine-inlet temperatures above 1300°C, demonstrating the importance of gas-turbine cost on future developments.

A thermoeconomic analysis of a cogeneration system consisting of a gas turbine linked to a waste boiler and situated in an institutional setting was reported by Luz-Silveira et al. (2002). The cogeneration system provided approximately one third of the 9-MW electrical demand as well as 1.8 kg/s of saturated steam (at 0.86 MPa) from a single fuel source. The study focused on equipment replacement. Based on pay-back period and maximum savings in 10 years, the best systems were found to be those using the Kawasaki Heavy Industries M1T-06 gas turbine and the Hitachi Zosen CCS7 gas turbine, respectively.

Colpan and Yesin (2006) carried out a thermoeconomic comparison using the specific-exergy costing method of several configurations of combined-cycle cogeneration systems, considering a varying steam demand. Cost balances were applied to the sub-systems of each cogeneration system and unit exergy costs found. Combined-cycle cogeneration systems with condensing steam turbines that permit steam extraction were determined on a thermoeconomic basis to be advantageous to those with back-pressure steam turbines.

For gas turbine-based plants for power and desalination, thermoeconomic analyses were carried out by Rensonnet et al. (2007), and exergy costs for such a plant were determined by Uche et al. (2006).

Methods for extending exergy accounting and thermoeconomics with environmental factors were applied to a cogeneration based on a gas turbine, yielding an optimal design (Sciubba, 2001, 2003).

5.3. Heating

Thermoeconomics was applied by Ozturk et al. (2006) to optimize hot water piping systems. Several approaches were considered, including: sequential optimization of pipe diameter based on minimization of total cost without considering heat losses, sequential optimization of insulation thickness based on minimization of cost of insulation and heat losses, and simultaneous optimization of pipe diameter and insulation thickness based on maximization of exergy efficiency and cost minimization.

A greenhouse heating system using a solar-assisted ground-source heat pump was assessed using EXCEM analysis and by investigating the relation between capital costs and thermodynamic losses for devices (Ozgener and Hepbasli, 2005). Ratios of thermodynamic loss rate to capital cost were determined, demonstrating that devices in successful air conditioning are configured so as to achieve an overall optimal design by appropriately balancing the exergy-based and economic characteristics of the overall system and its devices.

Exergoeconomic analyses have been reported of ground-source heat pump systems for building applications (Ozgener et al., 2005) and of geothermal district heating systems (Ozgener et al., 2007).

5.4. Refrigeration and Air Conditioning

An exergoeconomic optimisation was reported by Sahoo et al. (2004) of an aqua-ammonia absorption refrigeration system for air conditioning. The authors minimized the overall operation and amortization cost and utilized exergetic costs. A significant product cost reduction (15%) was observed for a small capital cost increase (2%).

An exergoeconomic optimisation of the geometry of continuous fins on an array of tubes of a refrigeration air-cooled condenser was performed by Jassim et al. (2005). The exergy destruction was investigated considering changes in geometry by varying condenser tube diameter (which also cause condenser tube length and total heat transfer area to vary) for a selected condenser capacity. An optimum heat exchanger area was found by developing an optimum balance between the components of exergy destruction. The total cost was expressed based on annual capital cost and compensation cost for irreversibilities.

Jassim et al. (2005) also carried out a thermoeconomic optimization of the geometry of a precooling air reheater for air conditioning. The total cost function was expressed on an annualized basis as the sum of the precooler capital cost and the operating cost attributable to the precooler irreversibility. This total cost function was optimized based on the optimum heat transfer area and the total irreversibilities.

To investigate the effect of matrix thermal conductivity on the performance of an air-conditioning rotary regenerator, Jassim and Khir (2004) performed an exergoeconomic optimization. The optimum geometry of the rotary regenerator was determined using the unit exergy cost of the warm air delivered as the objective function. The operating cost was determined using different unit costs for the pressure and thermal components of exergy. Examining the effect of varying the regenerator geometry revealed that the effect of thermal conductivity on the minimum exergoeconomic objective function is slightly over 1% for mild steel and 3% for aluminum.

An exergoeconomic analysis of a cold thermal energy storage system using a glycol working fluid has been performed by Bakan et al. (2008).

5.5. Chemical and Fuel Processing and Separation

Processes for chemical and fuel processing and separation vary widely. To improve designs, some of these have been investigated with exergy-based economic methods. For example, as noted earlier, desalination processes were examined with thermoeconomics (Rensonnet et al., 2007; Uche et al., 2006), and the analysis method based on exergetic, economic, environmental and other parameters proposed by Tonon et al. (2006) was applied to bioethanol production.

A thermoeconomic investigation was carried out by Nafey et al. (2006) of a multi-effect evaporation system and a combined multi-effect evaporation and multi-stage flash system. The assessment showed that, for the combined system, the operating cost decreases with increasing module number, although the capital cost increases. The minimum unit product cost was found to be $\$1.65/m^3$ for a system with eight modules. The combined system was found to be more economic than stand-alone systems.

Boyarskii et al. (1992) carried out thermal-economic analysis of obtaining high-purity products from air separators.

5.6. Others

EXCEM analysis has been applied to a range of thermal systems and processes by Rosen and Dincer (2003c).

6. Illustrative Examples

Examples are presented of applications of exergy analysis and several exergy-based economic methods to illustrate the concepts and methods discussed in this chapter.

6.1. Exergy Analysis

An exergy analysis is performed of coal-fired power generation to demonstrate how exergy analysis improves understanding of the thermodynamic performance of processes compared with energy analysis, and identifies areas with significant potential for improvement.

The Nanticoke Generating Station, a typical coal-fired power plant, is considered (Ontario Hydro, 1996). Located in Ontario, Canada, it has been operated since 1981 by the provincial utility, Ontario Power Generation (formerly Ontario Hydro).

6.1.1. Process Description

Each unit in the coal-fired power plant has a net electrical output of approximately 500 MW. A flow diagram for a single unit is shown in Figure 2. Symbols identifying the flows are described in Table 2a for material flows and in Table 2b for electrical and thermal flows. The main process data are listed in Table 3. The plant is described below, based on the four main sections identified in Figure 2:

- *Preheating:* The temperature of the feedwater is increased in several heaters and the pressure is increased in several pumps, to design levels for entering the steam generator.
- *Steam generation:* Water is converted to steam using combustion. Eight pulverized-coal-fired steam generators each produce 453.6 kg/s steam at 16.89 MPa and 538°C,

and 411.3 kg/s of reheat steam at 4.00 MPa and 538°C. Air is supplied to the furnace by two motor-driven forced draft fans, and regenerative air preheating is used. After treatment in an electrostatic precipitator, the flue gas exits through two multi-flued, 198 m-high stacks.

- *Power production:* Steam from the steam generation section passes through a series of turbine generators, attached to a transformer. The turbine generator contains one single-flow high-pressure cylinder, one double-flow intermediate-pressure cylinder and two double-flow low-pressure cylinders. Steam from the high-pressure cylinder is reheated in the combustor, and several steam flows for feedwater heating are extracted.
- *Condensation:* The low-pressure turbines exhaust at 5 kPa to the condenser, where the steam is condensed with cooling water from Lake Erie, which is restricted to a specified temperature rise.

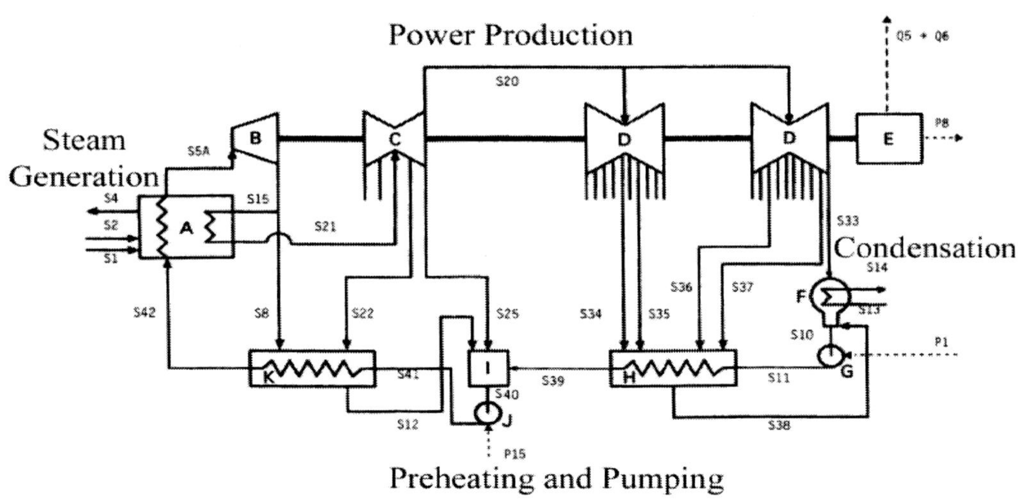

Figure 2. Single unit of the coal-fired power plant. For the devices, A: steam generator and reheater, B: high-pressure turbine, C: intermediate-pressure turbine, D: low-pressure turbines, E: generator and transformer, F: condenser, G: hot well pump, H: low-pressure heat exchangers, I: open deaerating heat exchanger, J: boiler feed pump, K: high-pressure heat exchangers. Symbols identifying flows are explained in Table 2. Lines exiting turbines represent flows of extraction steam.

Table 2a. Data for material flows in coal-fired power plant in Figure 2

Flow	Intensive flow properties			Flow rates		
	Temperature (°C)	Pressure (MPa)	Vapor fraction*	Mass (kg/s)	Energy (MW)	Exergy (MW)
H_2O flows						
S5A	538.00	16.2	1.0	453.59	1585.28	718.74
S8	323.36	3.65	1.0	42.84	135.44	51.81
S10	35.63	0.0045	0.0	367.85	36.52	1.20
S11	35.73	1.00	0.0	367.85	37.09	1.70

Table 2a. Continued

Flow	Intensive flow properties			Flow rates		
	Temperature (°C)	Pressure (MPa)	Vapor fraction*	Mass (kg/s)	Energy (MW)	Exergy (MW)
S12	188.33	1.21	0.0	58.82	50.28	11.11
S13	15.00	0.101	0.0	18,636	0.00	0.00
S14	23.30	0.101	0.0	18,636	745.95	10.54
S15	323.36	3.65	1.0	410.75	1298.59	496.81
S20	360.50	1.03	1.0	367.85	1211.05	411.16
S21	538.00	4.00	1.0	410.75	1494.16	616.42
S22	423.23	1.72	1.0	15.98	54.54	20.02
S25	360.50	1.03	1.0	26.92	88.64	30.09
S33	35.63	0.0045	0.93	309.62	774.70	54.07
S34	253.22	0.379	1.0	10.47	32.31	9.24
S35	209.93	0.241	1.0	23.88	71.73	18.82
S36	108.32	0.0689	1.0	12.72	35.77	7.12
S37	60.47	0.0345	1.0	11.16	30.40	5.03
S38	55.56	0.0133	0.0	58.23	11.37	0.73
S39	124.86	1.00	0.0	367.85	195.94	30.41
S40	165.86	1.00	0.0	453.59	334.86	66.52
S41	169.28	16.2	0.0	453.59	347.05	77.57
S42	228.24	16.2	0.0	453.59	486.75	131.93
*Other material flows (fuel, air, combustion gases)***						
S1	15.00	0.101	solid	41.74	1367.58	1426.73
S2	15.00	0.101	1.0	668.41	0.00	0.00
S3***	1673.59	0.101	1.0	710.15	1368.00	982.85
S4	119.44	0.101	1.0	710.15	74.39	62.27

* Vapour fraction is listed as 0 for liquids and 1 for saturated or superheated vapours or gases.

** By volume, S1 is modeled as 100% carbon, S2 as 79% N_2 and 21% O_2, and S3 and S4 as 79% N_2, 6% O_2 and 15% CO_2.

*** Flow S3 (not shown in Figure 2) represents the hypothetical hot product gases for adiabatic combustion.

Table 2b. Data for electrical and thermal flows in coal-fired power plant in Figure 2

Flow	Energy flow rate (MW)	Exergy flow rate (MW)
Thermal flows		
Q5	5.34	0.00
Q6	5.29	0.00
Electrical flows		
P1	0.57	0.57
P8	523.68	523.68
P15	12.19	12.19

Table 3. Main process data for full-load operation of a coal-fired power plant

Quantity	Value
Flow rates	
Mass (kg/s)	
Primary steam	454
Reheat steam	411
Coal	47.9
Cooling water	18,636
Electricity (MW)	
Gross output	524
Internal use	19
Net output	505
Intensive properties	
Temperature (°C)	
Boiler feedwater	253
Primary steam	538
Reheat steam	538
Flue gas	120
Cooling water rise	8.3
Pressure (MPa)	
Primary steam	16.9
Reheat steam	4.0
Condenser	0.005

6.1.2. Analysis

Comprehensive energy and exergy analyses are performed. Base enthalpy and chemical exergy values reported elsewhere (Rosen and Dincer, 2003a) are used.

Several simplifications are applied:

- coal is modeled as graphite and air as 79% nitrogen and 21% oxygen by volume,
- the turbine isentropic and mechanical efficiencies are 80% and 95%, respectively, and
- the generator and transformer efficiencies are both 99%, with heat losses from their surfaces occurring at 15°C.

The overall energy efficiency η is evaluated as

$$\eta = \frac{\text{Net energy output with electricity}}{\text{Energy input}} \tag{23}$$

and the overall exergy efficiency ψ as

$$\psi = \frac{\text{Net exergy output with electricity}}{\text{Exergy input}} \tag{24}$$

For most plant components and sections, similar efficiency expressions are applied. But a "condenser efficiency" is not straightforwardly defined, as its purpose is not to generate a product, but rather to reject waste heat.

A common reference-environment model (Gaggioli and Petit, 1977; Rodriguez, 1980) is employed in evaluating energy and exergy, but with a temperature set at the approximate mean for the lake cooling water or 15°C. The reference-environment pressure is 1 atm, and the chemical composition is air saturated with water vapor and the following condensed phases at 15°C and 1 atm: water, gypsum and limestone.

6.1.3. Energy and Exergy Flows and Efficiencies

Energy and exergy flow rates are given in Table 2a for material flows and in Table 2b for electrical and thermal flows. Device exergy consumptions are provided in Table 4. Figures 3a and 3b illustrate the net energy and exergy flows and exergy consumptions for the four main process sections.

Table 4. Exergy consumption rates for coal-fired power plant

Section	Device	Exergy consumption rate (MW)
Steam generation	Steam generator (including combustor)	659.0
Power production	Turbines	107.9
	Generator and transformer	10.6
	Section total	118.5
Condensation	Condenser	43.1
Preheating	Heat exchangers	22.2
	Pumps	1.2
	Section total	23.4
Overall		844.0

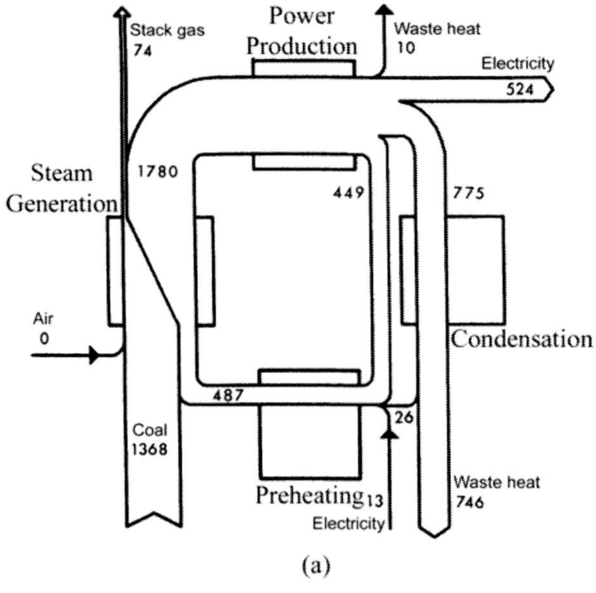

(a)

Figure 3. Continued on next page.

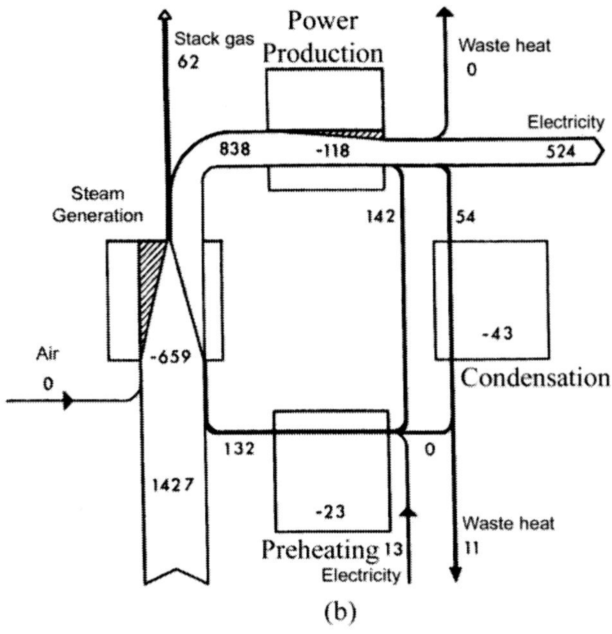

Figure 3. Flow rates (in MW) through the four main sections of the coal-fired power plant of a) energy and b) exergy. Thicknesses of flow lines are proportional to the magnitude of the flow rate of the relevant commodity.

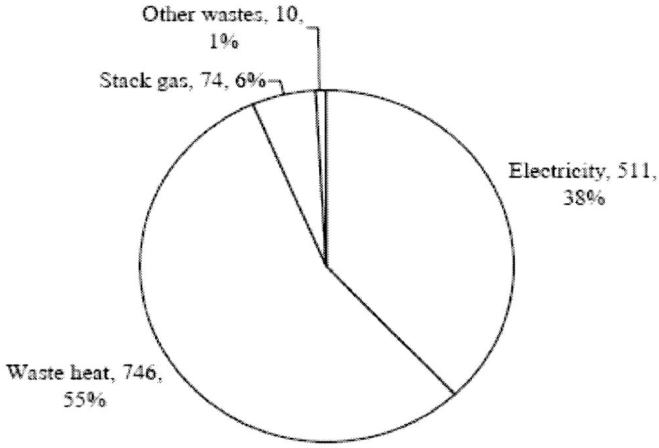

Figure 4. Distribution of energy outputs for a coal-fired power plant, showing values in megawatts and as a percentage of the total output.

The distribution of outputs (the electrical product as well as wastes) of energy is shown in Figure 4, and of exergy is shown in Figure 5. The latter figure also shows the loss associated with overall exergy consumption. The overall exergy consumption is broken down in Figure 6.

It is noted that 1) energy and exergy values and efficiencies for the overall process and its sections are in agreement with the literature for similar processes, and 2) realistic variations in reference-environment properties do not significantly affect the energy and exergy results.

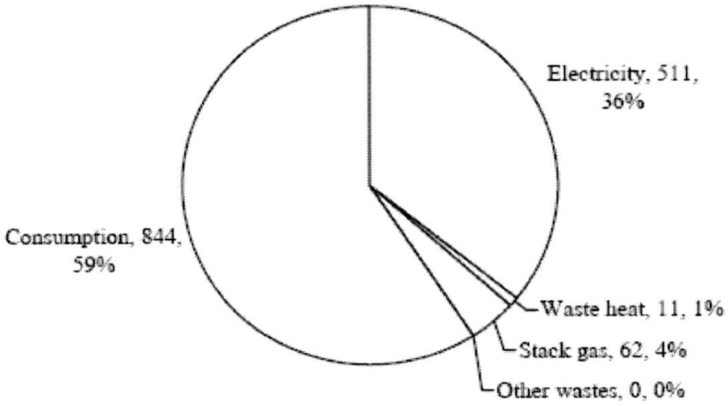

Figure 5. Distribution of exergy outputs and exergy consumption for a coal-fired power plant, showing values in megawatts and as a percentage of the total output and consumption.

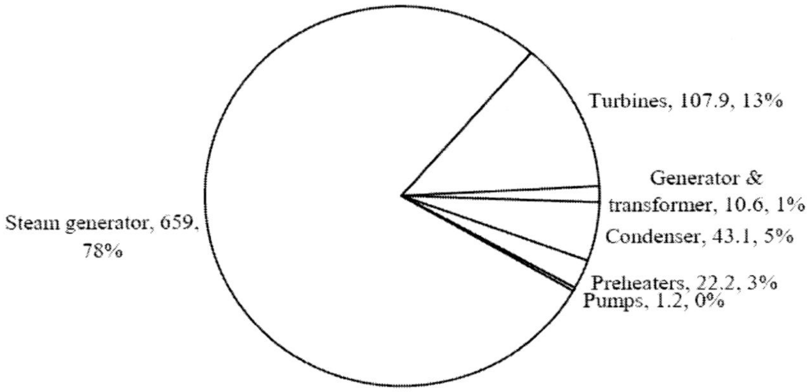

Figure 6. Breakdown by device of the exergy consumptions for a coal-fired power plant, showing values in megawatts and as a percentage of the total exergy consumption.

Overall plant

Overall energy and exergy efficiencies, where coal provides the only input energy or exergy, are

$$\eta = \frac{(524 - 13)\,\text{MW}}{1368\,\text{MW}}(100\%) = 37\%$$

and

$$\psi = \frac{(524 - 13)\,\text{MW}}{1427\,\text{MW}}(100\%) = 36\%$$

The difference in efficiencies is minor, and attributable to the specific chemical exergy of coal being slightly greater than its specific base enthalpy.

Preheating section

Energy losses are minor in the preheating section (under 10 MW), as are exergy losses (about 20 MW). Internal consumptions comprise most of the exergy losses.

Steam generation section

Energy and exergy efficiencies for the steam generation section, considering the increase in energy or exergy of the water as the product, are

$$\eta = \frac{[(1585 - 847) + (1494 - 1299)]\ \text{MW}}{1368\ \text{MW}}(100\%) = 95\%$$

and

$$\psi = \frac{[(719 - 132) + (616 - 497)]\ \text{MW}}{1427\ \text{MW}}(100\%) = 49\%$$

The steam generation section is significantly more efficient based on energy than exergy, implying that although 95% of the input energy preheats water, the energy is degraded as it is transferred. Exergy analysis highlights this degradation.

Exergy consumptions in the steam generation section are substantial, accounting for 659 MW or 72% of the 916 MW total exergy loss. Of the 659 MW exergy consumption rate in this section, 444 MW is due to combustion and 215 MW to heat transfer.

Power production section

In the power production section, energy losses are small (less than 10 MW total), and exergy losses are moderately small (around 150 MW). The exergy losses are mainly internal consumptions.

Condensation section

In the condensers,

- a large quantity of energy enters (775 MW), of which close to 100% is rejected; and
- a small quantity of exergy enters (54 MW), of which about 25% is rejected and 75% is internally consumed.

Thus, energy results erroneously suggest that almost all losses in electricity-generation potential are associated with condenser heat rejection, while exergy analyses demonstrate quantitatively and directly that the condensers are responsible for little of these losses (Figures 4-6). This discrepancy arises because condenser heat is rejected nearly at the environment temperature.

Exergy results show that the condenser waste is relatively insignificant for the process. The exergy rejected by the condensers is less than 4% of the net exergy produced by the plant, while the energy rejected exceeds the net energy produced by approximately 50%.

6.1.4. Further Discussion

Several illuminating insights have been attained about coal-fired power generation:

- Energy losses are mainly in emissions (i.e., condenser heat rejection), and exergy losses are primarily with consumptions (mainly in steam generation) and somewhat with cooling water and stack gases.
- Although overall energy and exergy efficiencies are similar, energy analyses do not identify the locations and causes of inefficiencies, while exergy analyses do so systematically.
- Since devices with the largest thermodynamic losses have the greatest margins for efficiency improvement, efforts to increase the efficiencies of coal-fired power generation should focus on the combustor. Technologies that generate electricity without combustion (e.g., fuel cells) or utilize heat at high temperatures could increase efficiencies significantly.
- The use of heat rejected by condensers only increases the exergy efficiencies by a few per cent. Cogeneration systems, which produce heat at useful temperatures at the expense of reduced electrical output, can have greater efficiencies than conventional power generation, but the merit of cogeneration systems must be determined using exergy because energy analyses tend to overstate performance.

6.2. EXCEM Analysis

The application of EXCEM analysis is considered for a coal-fired power plant and for a cogeneration plant for electricity and heat.

6.2.1. EXCEM Analysis of a Coal-Fired Power Plant

The typical coal-fired power plant considered previously (Figure 2) is investigated using EXCEM analysis. The analysis of a complex system like a power plant can be carried out for the overall system only, or as a set of individual EXCEM analyses of the devices constituting the overall system.

Figure 7 illustrates the results of an EXCEM analysis of the overall coal-fired power plant. In this figure, the input and generation costs are allocated to the product electricity, and the exhaust cooling water and stack gas are treated as wastes.

The costs in Figure 7 are in 2007 Canadian dollars. These costs are evaluated by modifying the original data (Ontario Hydro. 1969, 1973, 1991, 1983, 1996), in 1982 Canadian dollars, using the Consumer Price Index (CPI) tabulated by Statistics Canada (accessible via www.statcan.ca). The CPI data, which represent changes in prices of all goods and services purchased for consumption, indicate that $1.00 in 1982 has the same buying power as $2.04 in 2007.

Note that, in this example, subjective decisions have been made regarding the allocation of costs among outputs, which include net electricity, stack gas and cooling water. In principle, each may each be allocated part of the input and generation costs, depending on its use. The exhaust cooling water may be treated as a by-product rather than a waste, in which case part of the total costs would be allocated to it.

6.2.2. EXCEM Analysis of a Cogeneration Plant

Plant description

The cogeneration plant considered operates at steady-state and produces 17.9 kg/s of steam at a pressure of 44.2 atm and a temperature of 399°C in the boiler. After passing through the turbine generator, the steam is exhausted at 0.065 atm and 37.8°C. Data for the illustrative example are based on a previous investigation of a steady-state cogeneration plant (Reistad and Gaggioli, 1980).

Analysis and results

The reference environment used in the energy and exergy evaluations has a temperature of 10°C, a pressure of 1 atm and a composition as described in the reference-environment model of Gaggioli and Petit (1977).

Values of cost generation rate are evaluated as follows:

$$\text{Cost generation rate} = (\text{Capital cost} \times \text{Ammortization factor})/\text{Load factor} \quad (25)$$

where the ammortization factor is 0.08 and the load factor is 0.7. The total costs of a device are spread over its life with the ammortization factor, accounting for the time value of money.

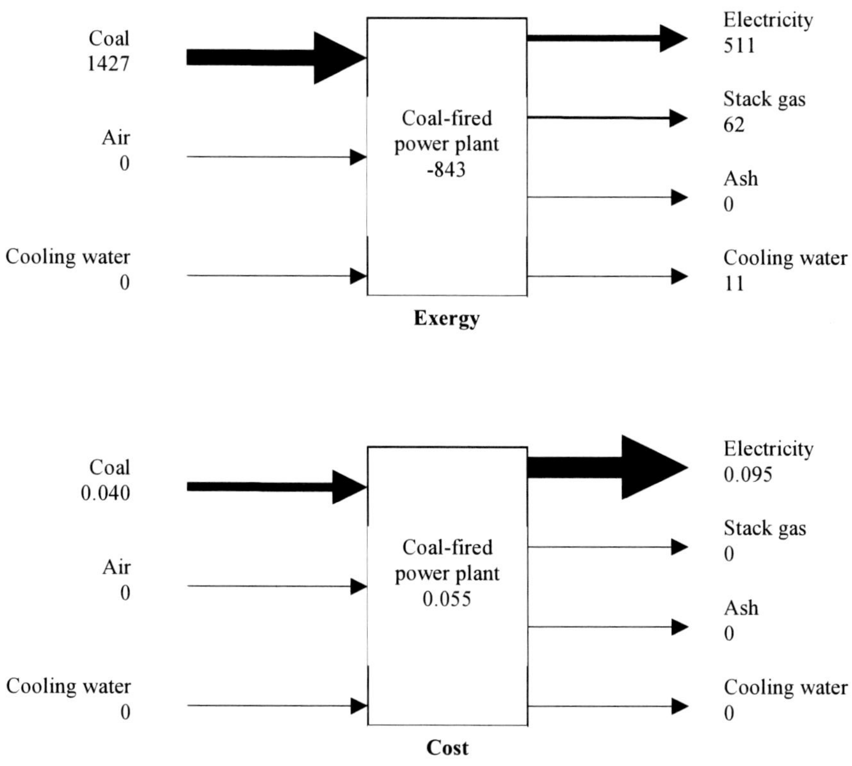

Figure 7. Continued on next page.

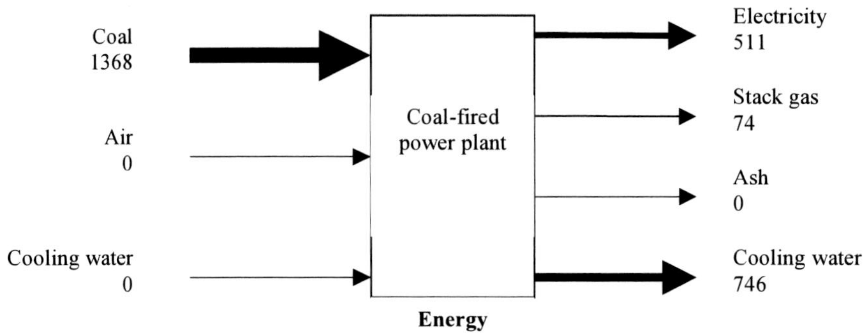

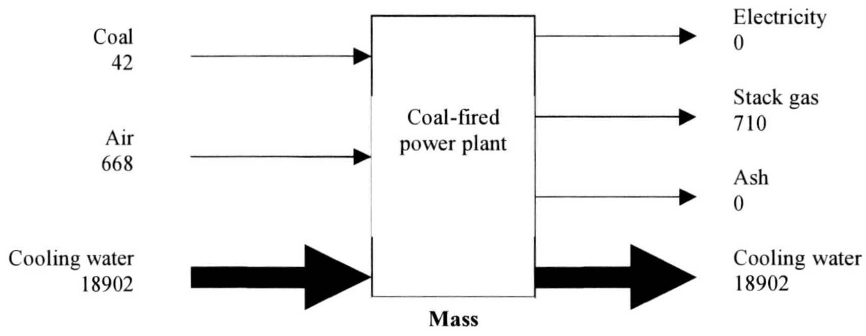

Figure 7. Balances of exergy, cost, energy and mass for a coal-fired power plant. The exergy balance shows flow rates and consumption rate (negative value in plant) of exergy (in MW). The cost balance shows flow rates and creation rate (value in plant) of cost (in 2007 Canadian $/kWh). The energy balance shows energy flow rates (in MW) and the mass balance mass flow rates (in kg/s). In all balances, thicknesses of flow lines are proportional to the magnitude of the flow rate of the relevant commodity.

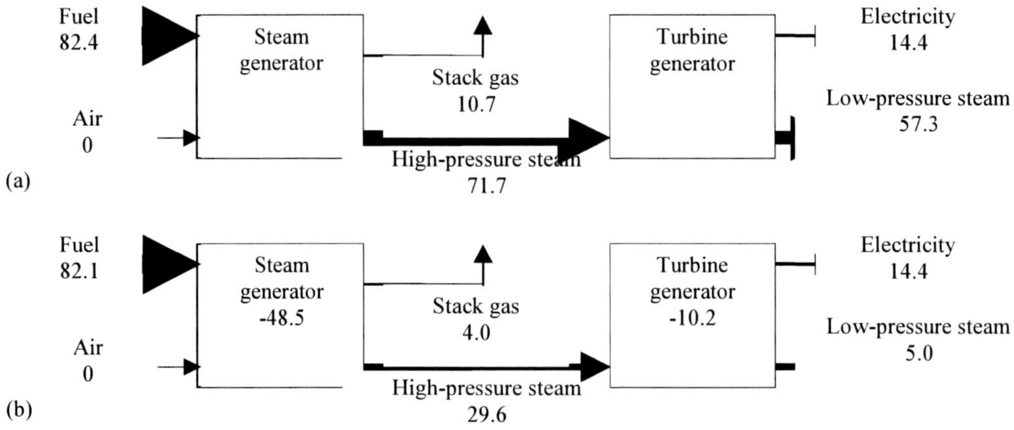

Figure 8. Continued on next page.

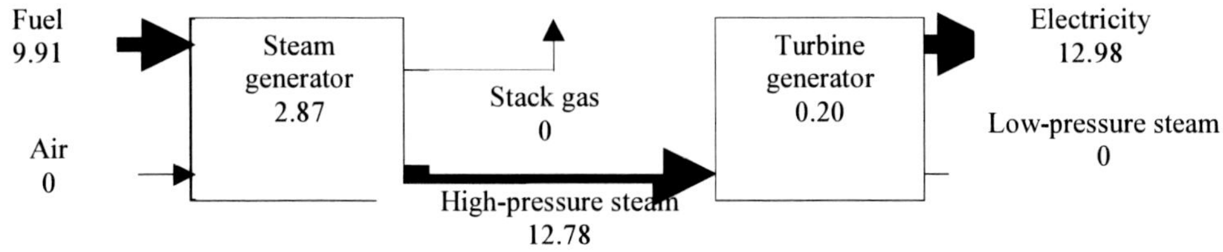

(c)

Figure 8. Flow rates of three commodities through a cogeneration plant for the production of electricity and low-pressure steam: (a) energy (in MW), (b) exergy (in MW) and (c) cost (in 2007 US M$/yr). Rates of exergy consumption and cost creation are given within the devices in (b) and (c), respectively. Thicknesses of flow lines are proportional to the magnitude of the flow rate of the relevant commodity.

Flow rates of exergy, energy and cost are illustrated in Figures 8a, 8b and 8c, respectively, as are rates of exergy consumption and cost creation. Inputs to the boiler of feedwater, for which the associated flow rates of energy, exergy and cost are approximately zero, are omitted from the diagram. Costs have been adjusted from those originally reported to 2007 US dollars using the US Consumer Price Index.

Further analysis and discussion

The relations involving cost and exergy, which perhaps present the most informative results in an EXCEM analysis, are examined in detail. The flow rates of exergy and cost at different points in the cogeneration process are shown in Figure 9. A monotonically decreasing composite line is traced in all cases.

The data in Figure 9 are modified to highlight the exergy losses and their relations to cost creation. In Figure 10, exergy loss rates (total and internal) are shown as a function of cost creation rates for the two devices in the process: the boiler and the turbine generator.

The data are modified further in Figure 11, where cumulative exergy loss rates (total and internal) at different points in the plant are illustrated as a function of cumulative cost creation rate.

It is noted that the intensive properties of the reference environment and the costs associated with inputs are required to determine the points for Figure 9, but not for Figures 10 or 11. Figure 9 is consequently more demanding to generate.

Several observations can be made regarding Figures 9-11:

- All three figures are structured to demonstrate in different ways the general trade off that is typically made between cost and efficiency (or inefficiency). In the extremes, performance approaches ideality as the investment made becomes increasingly large, while the total exergy input is wasted as the investment made becomes increasingly small. In real systems, a balance is obtained between exergy loss and cost creation.
- The figures demonstrate that exergy and cost are the only EXCEM quantities subject to non-conservation laws. With increasing cost creation, the exergy curves in Figure 9 always decrease and in Figures 10 and 11 always increase because the values of cost creation and exergy loss must be positive for any device.
- The slopes and magnitudes of the individual lines in the figures indicate characteristics of the corresponding devices.

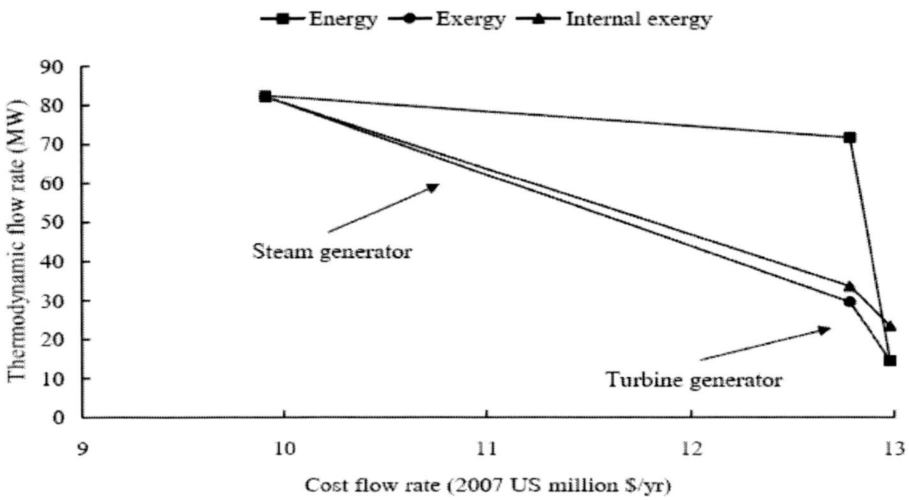

Figure 9. Variation with cost flow rate of energy, exergy and exergy consumption rates for devices in a cogeneration plant.

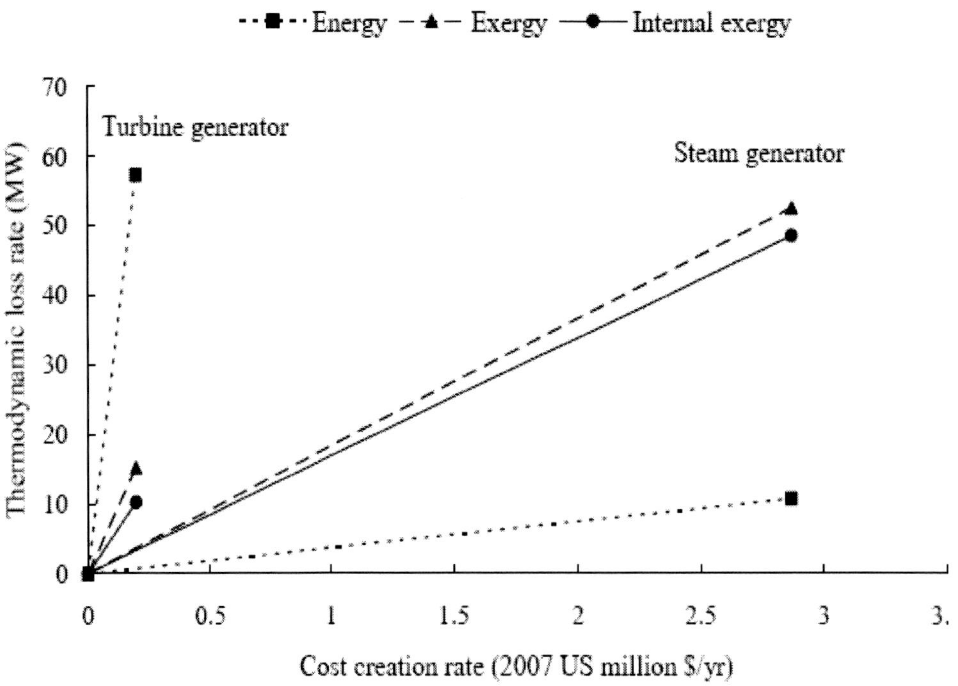

Figure 10. Variation with cost creation rate of rates of energy loss, exergy loss and exergy consumption for devices in a cogeneration plant.

For comparative purposes, energy values are also shown in the figures. Specifically, energy flow rates are shown in Figure 9, and energy loss rates in Figures 10 and 11. Since "internal" energy losses do not exist due to the principle of energy conservation, only total energy losses are shown in the plots. Two key observations follow:

- The curves representing energy in Figures 9-11 are somewhat similar to the exergy curves, but they are misleading because they treat all energy losses equally. Different forms of energy are not of equal value because some can be completely converted into other energy forms while others can not, even in an ideal process. Different exergy forms are normalized to a common basis; an exergy quantity can be converted into any other form of exergy of the same amount in an ideal process.
- Comparisons of Figures 9-11 reinforce the concepts embedded in exergy-based pricing which suggest, since exergy is often a consistent measure of a commodity's value but energy is in general not a consistent measure of value, that costing should be based on exergy rather than energy.

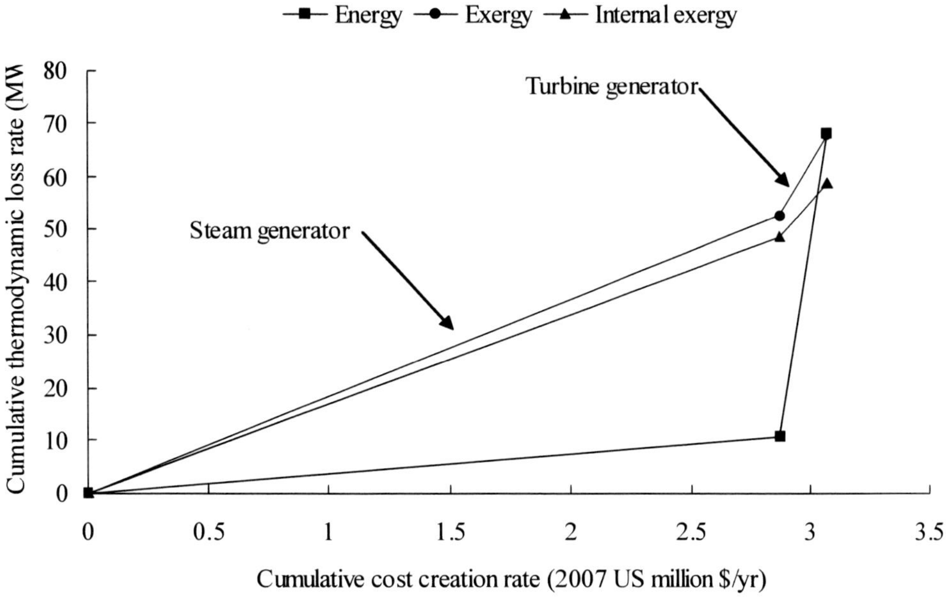

Figure 11. Variation with cumulative cost creation rate of cumulative rates of energy loss, exergy loss and exergy consumption for devices in a cogeneration plant.

6.3. Loss-Cost Ratio Analysis

Loss-cost ratio analysis is illustrated for three types of thermal power generation, two representing fossil fuels (coal and oil) and one representing a non-fossil resource (nuclear). These are illustrated and contrasted in Figure 12.

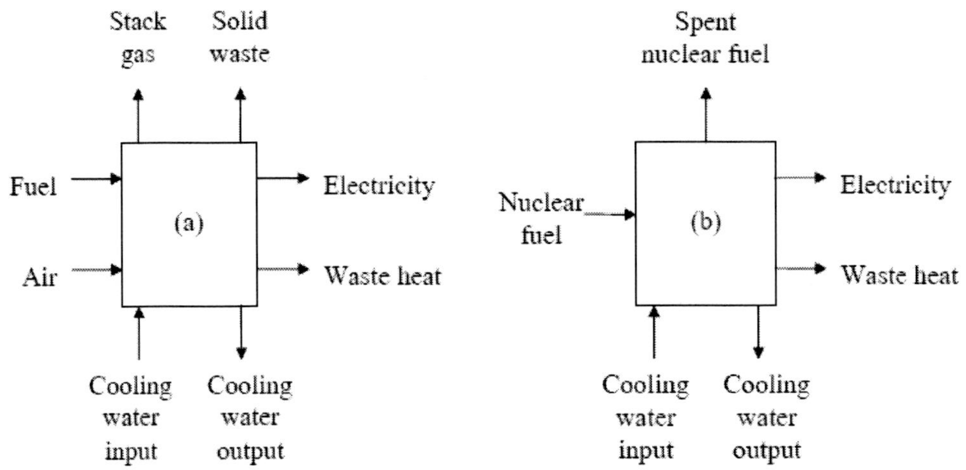

Figure 12. Schematic diagrams for two categories of thermal power plants: (a) fossil fuel, (b) nuclear.

6.3.1. Descriptions of the Thermal Power Plants

The economic and thermodynamic data used in the analyses and the corresponding sources of the data are given. Capital cost is for simplicity the only economic term considered here. As noted earlier, this approach is often a reasonable simplification. Table 5 presents selected primary characteristics of the four stations. The following thermodynamic losses are provided for all three types of power generation: exergy, exergy consumption, waste exergy emission and energy. All losses are associated with these quantities since the stations normally operate at steady state.

Coal

Coal-fired power generation is represented by the Nanticoke generating plant considered in the illustrative example of exergy analysis.

The overall plant (including the devices examined separately below and all other plant devices) is considered, and it is subdivided into the following devices:

- turbines,
- steam generators (including the steam generator and reheater, and the air preheater and fan),
- preheating devices (including the deaerator and the high- and low-pressure preheaters)
- pumps (such as condensate and boiler feed pumps), and
- condensers.

The analysis requires that thermodynamic losses be identified. Some losses can be identified straightforwardly. In coal-fired power generation, waste emissions include spent cooling water from the condenser, stack gas exiting the steam generator, and miscellaneous heat losses from all devices. But, ambiguity exists regarding what are the material waste emissions for the turbine generators and preheating devices. Consequently,

- three loss cases are considered for the turbine generators: none of the steam exhausted or extracted from the turbines is a waste (Case I), only the steam exhausted to the condenser from the low-pressure turbine is a waste (Case II) and all steam exhausted or extracted from the turbines is a waste (Case III).
- two loss cases are considered for the preheating devices: no exiting flows are wastes (Case I) and the flow from the preheating devices to the condenser is a waste (Case II).

Table 5. Selected information for three types of thermal power generation

Type of thermal power generation			Efficiency (%)	
Generation type	Representative plant	Net electrical output (MW)	Energy	Exergy
Coal	Nanticoke	511	37.4	35.8
Oil	Lennox	512	37.0	34.8
Nuclear	Bruce B	842	31.0	31.0

Source: (Rosen, 1990, 1991).

For the coal-fired plant, thermodynamic data have been given earlier and thermodynamic losses are shown in the top section of Table 6 for the breakdown given above.

The costs in Table 6 are in 2007 Canadian dollars. These costs are evaluated by modifying the original data using the Consumer Price Index. The original costs (Ontario Hydro. 1969, 1973, 1991, 1983, 1996) for the Nanticoke station are in 1982 U.S. dollars. The 1982 U.S.-Canada exchange rate is used to convert the costs to Canadian dollars. The CPI and exchange-rate data indicate that $1.00 (U.S.) in 1982 has the same buying power as $2.54 (Canadian) in 2007.

The economic data for the coal power plant have been compared with data for typical coal power plants of similar size, such as the coal-fired Harry Allen Station in southern Nevada containing three 500-MWe units (Tsatsaronis and Winhold, 1985) and seen to be consistent.

Oil

Oil-fired power generation is represented by the Lennox generating plant in Ontario, which has been examined using exergy elsewhere (Rosen and Dincer, 2003b). The analysis utilized data from Ontario Power Generation, formerly Ontario Hydro (1979).

The overall plant is considered, and it is broken down for analysis following the breakdown presented above for coal-fired power generation. Two additional device breakdowns are considered for the oil-fired plant: the gland seal condenser and the low-pressure preheaters with the gland seal condenser. In both instances, two cases are considered: no material wastes exit, and flows to the condenser are wastes.

For the oil-fired plant, thermodynamic losses and capital costs are shown in the middle section of Table 6 for the breakdown given above.

The original costs for the Lennox station are in 1973 Canadian dollars, and were modified to 2007 Canadian dollars following the method outlined above for coal power generation. The CPI data indicate that $1.00 (Canadian) in 1973 has the same buying power as $4.64 (Canadian) in 2007.

Table 6. Device parameter values for coal, oil and nuclear power plants (using 2007 Can. $)

Plant and section/device	Case	Cost, K (10^6 $)	Thermodynamic loss rate (MW)				Loss-to-cost ratio (W/$)			
			L_{en}	L_{ex}	$L_{ex\text{-}i}$	$L_{ex\text{-}e}$	R_{en}	R_{ex}	$R_{ex\text{-}i}$	$R_{ex\text{-}e}$
Coal-fired plant (overall)		335.35	918.400	1025.920	944.160	81.760	2.744	3.058	2.811	0.246
Steam generators		188.05	82.880	806.400	736.960	69.440	0.437	4.290	3.920	0.370
Turbines	I	78.09	0.000	120.960	120.960	0.000	0.000	1.546	1.546	0.000
	II	78.09	868.000	181.440	120.960	60.480	11.144	2.318	1.546	0.773
	III	78.09	1370.880	340.480	120.960	219.520	17.550	4.368	1.546	2.811
Preheating section	I	19.56	0.000	25.760	25.760	0.000	0.000	1.322	1.322	0.000
	II	19.56	12.768	26.902	25.760	1.344	0.650	1.389	1.322	0.067
Condenser		10.61	835.520	59.360	48.160	11.200	78.870	5.622	4.536	1.053
HP preheater	I	3.91	0.000	7.123	7.123	0.000	0.000	1.826	1.826	0.000
	II	3.91	56.336	19.600	7.123	12.432	14.437	1.826	1.826	3.181
LP preheater	I	2.68	0.000	12.096	12.096	0.000	0.000	4.514	4.514	0.000
	II	2.68	12.768	12.880	12.096	0.818	4.760	4.514	4.514	0.302
Deaerator		0.64	0.000	0.627	0.627	0.000	0.000	0.986	0.986	0.000
Condensate pumps		1.59	0.000	0.078	0.078	0.000	0.000	0.045	0.045	0.000
Oil-fired plant (overall)		192.9	968.822	1074.651	1001.302	73.349	5.029	5.578	5.186	0.381
Steam generators		60.80	232.467	917.549	862.109	55.440	3.819	15.086	14.179	0.907
Turbines	I	67.22	0.000	75.824	75.824	0.000	0.000	1.131	1.131	0.000
	II	67.22	758.184	120.579	75.824	44.755	11.278	1.792	1.131	0.661
	III	67.22	1218.504	268.072	75.824	192.248	18.122	3.987	1.131	2.856
Preheating section	I	5.37	0.000	20.216	20.216	0.000	0.000	3.763	3.763	0.000
	II	5.37	7.000	20.485	20.216	0.269	1.299	3.853	3.763	0.049
Condenser		3.13	736.254	44.274	26.398	17.875	235.480	14.157	8.445	5.723
HP preheater	I	1.77	0.000	5.667	5.667	0.000	0.000	3.192	3.192	0.000
	II	1.77	44.397	15.557	5.667	9.890	25.021	8.770	3.192	5.578
LP preheater	I	1.35	0.000	6.082	6.082	0.000	0.000	4.502	4.502	0.000
	II	1.35	6.854	6.350	6.082	0.269	5.074	4.693	4.502	0.202

Table 6. Continued

Plant and section/device	Case	Cost, K (10^6 \$)	Thermodynamic loss rate (MW)				Loss-to-cost ratio (W/\$)			
			L_{en}	L_{ex}	$L_{ex\text{-}i}$	$L_{ex\text{-}e}$	R_{en}	R_{ex}	$R_{ex\text{-}i}$	$R_{ex\text{-}e}$
LP preheater and	I	1.56	0.000	6.642	6.642	0.000	0.000	4.245	4.245	0.000
gland seal condenser	II	1.56	7.000	6.910	6.642	0.269	4.480	4.424	4.245	0.168
Gland seal condenser	I	0.20	0.000	0.560	0.560	0.000	0.000	2.822	2.822	0.000
	II	0.20	0.146	0.560	0.560	0.000	0.739	2.822	2.822	0.000
Deaerator		0.51	0.000	3.326	3.326	0.000	0.000	6.563	6.563	0.000
Condensate pumps		0.59	0.000	0.011	0.011	0.000	0.000	0.019	0.019	0.000
Boiler feed pumps		0.93	0.000	4.570	4.570	0.000	0.000	4.917	4.917	0.000
Nuclear plant (overall)		391.3	1981.134	2100.762	2073.400	27.362	5.062	5.365	5.298	0.067
Steam generators		62.4	164.640	1764.538	1762.264	2.274	2.643	28.280	28.246	0.034
Turbines	I	261.5	0.000	46.110	46.110	0.000	0.000	0.179	0.179	0.000
	II	261.5	1848.056	155.030	46.110	108.920	7.067	0.594	0.179	0.414
	III	261.5	2755.379	378.907	46.110	332.797	10.573	1.445	0.179	1.277
Preheating section	I	13.63	0.000	28.168	28.168	0.000	0.000	2.061	2.061	0.000
	II	13.63	19.365	28.795	28.168	0.627	1.434	2.106	2.061	0.045
Condenser		5.76	1816.494	108.819	83.731	25.088	315.392	18.917	14.538	4.357
HP preheater	I	5.02	0.000	3.203	3.203	0.000	0.000	0.638	0.638	0.000
	II	5.02	145.006	30.722	3.203	27.418	28.862	6.093	0.638	5.454
LP preheater	I	3.36	0.000	19.891	19.891	0.000	0.000	5.914	5.914	0.000
	II	3.36	19.365	20.518	19.891	0.627	5.768	6.104	5.914	0.190
HP and LP preheaters	I	8.38	0.000	23.094	23.094	0.000	0.000	2.755	2.755	0.000
	II	8.38	164.371	51.139	23.094	28.045	19.611	6.104	2.755	3.349
Deaerator		1.82	0.000	1.523	1.523	0.000	0.000	0.840	0.840	0.000
Condensate pumps		1.54	0.000	0.448	0.448	0.000	0.000	0.291	0.291	0.000
Boiler feed pumps		1.89	0.000	2.509	2.509	0.000	0.000	1.333	1.333	0.000
Moisture separator	I	5.09	0.000	12.600	12.600	0.000	0.000	2.475	2.475	0.000
and heat exchanger	II	5.09	214.312	68.398	12.600	55.798	42.123	13.440	2.475	10.965

* LP and HP denote low-pressure and high-pressure, respectively.

Nuclear

Nuclear power generation is represented by the Bruce B generating plant in Ontario, which has been examined using exergy elsewhere (Rosen, 2001) using data from Ontario Power Generation, formerly Ontario Hydro (1969, 1985).

The plant breakdown presented above for coal-fired power generation is considered. In the nuclear plant, the steam generators include the nuclear reactor, the primary heat-transport-loop pump and the moderator cooler, rather than the air preheater, combustion chamber and fan. Additional device breakdowns, each with two cases, are considered for the nuclear plant:

- High- and low-pressure preheaters. Case I presumes no material wastes exit, while Case II takes flows from the low-pressure preheaters to the deaerator and from the high-pressure preheaters to the condenser to be wastes.
- Moisture separator and heat exchangers. Case I considers no material wastes exit, and case II takes flows from the moisture separator and heat exchanger to be wastes if they do not enter the low-pressure turbine.

For the nuclear power plant, thermodynamic losses and capital costs are shown in the bottom section of Table 6 for the breakdown given above.

The original costs for the Bruce B station are in 1976 Canadian dollars, and were modified to 2007 Canadian dollars following the method outlined above for coal power generation. The CPI data indicate that $1.00 (Canadian) in 1976 has the same buying power as $3.52 (Canadian) in 2007.

6.3.2. Analysis Results

For the thermodynamic loss rates and capital costs in Table 6 for coal-fired, oil-fired and nuclear power generation, the variation of four thermodynamic loss rates (L_{ex}, $L_{ex\text{-}i}$, $L_{ex\text{-}e}$ and L_{en}) with capital cost K are illustrated in Figures 13-16 for the overall plant and its devices. Energy loss rate and total, internal and external exergy loss rates are considered in Figures 13-16, respectively.

Values of the thermodynamic-loss-rate-to-capital-cost ratios R_{ex}, $R_{ex\text{-}i}$, $R_{ex\text{-}e}$ and R_{en} for each generating plant and its devices are listed in Table 6 and illustrated in Figure 17 for coal-fired power generation, Figure 18 for oil-fired power generation and Figure 19 for nuclear power generation. Note that the value of R for a device is given by the slope of the line in Figures 13-16 for the device.

Grouping all the device cases considered for a given generating plant, statistical data for the four thermodynamic-loss-rate-to-capital-cost ratios are determined (see Table 7). The statistical quantities examined include maximum (R_{max}), minimum (R_{min}), mean (R_m), standard deviation (SD(R)) and coefficient of variation (CV(R)). The mean R_m for a set of values, R_1, R_2, ... R_n, is a measure of the center of the set. The standard deviation SD(R) of the set is a measure of the absolute variation in the set of R values about the mean R_m. The coefficient of variation CV(R) is the standard deviation as a percentage of the mean. The latter measures the relative variation in a set of values about the mean and is independent of scale of measurement, permitting comparisons of the variations in several sets of data. In the present undertaking, all of the device groupings and cases are considered. But the values in Table 7 do not vary significantly if R values are based on either of the following alternative device groupings:

- all devices except the overall plant, or
- the devices which represent the most broken down components of each plant.

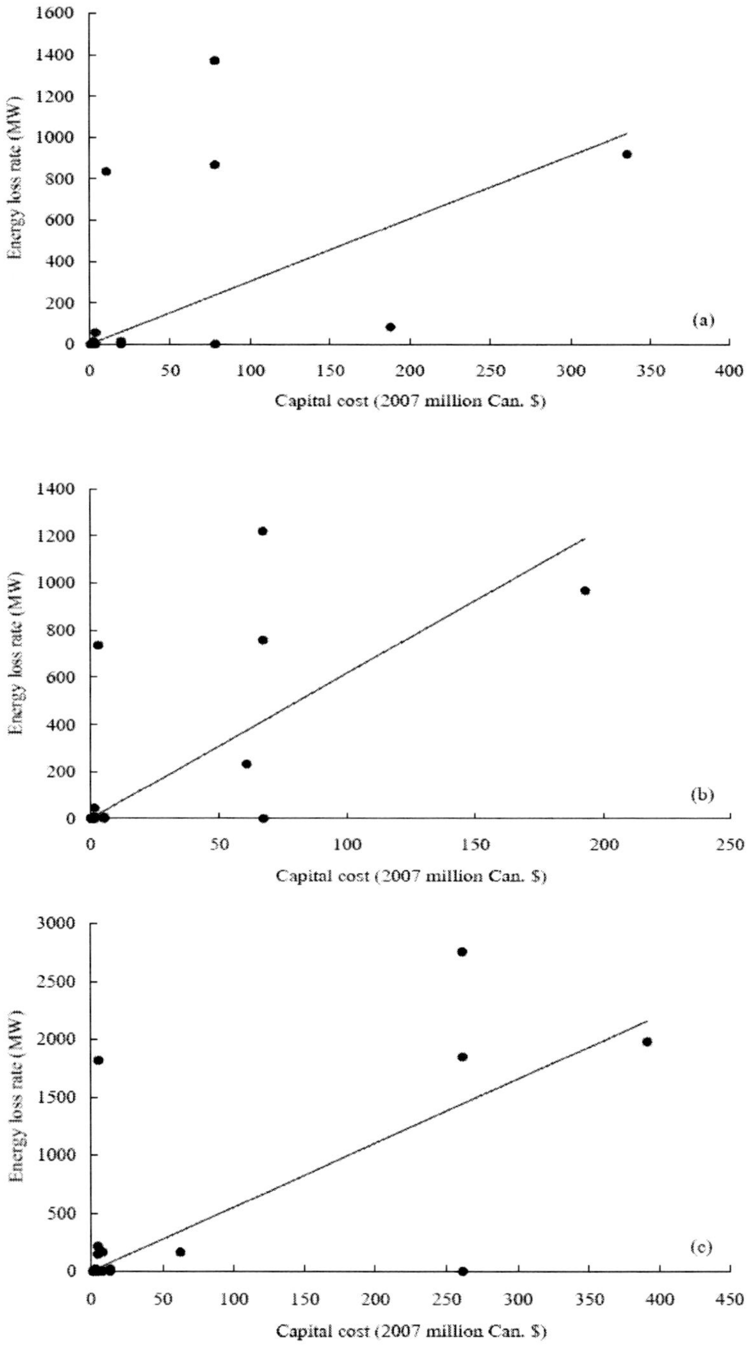

Figure 13. Energy loss rate as a function of capital cost for several devices and the overall plant for a) coal-fired power generation, b) oil-fired power generation and c) nuclear power generation. A trend line determined using linear regression and intersecting the origin is shown to help illustrate the degree to which data points are clustered around a common slope.

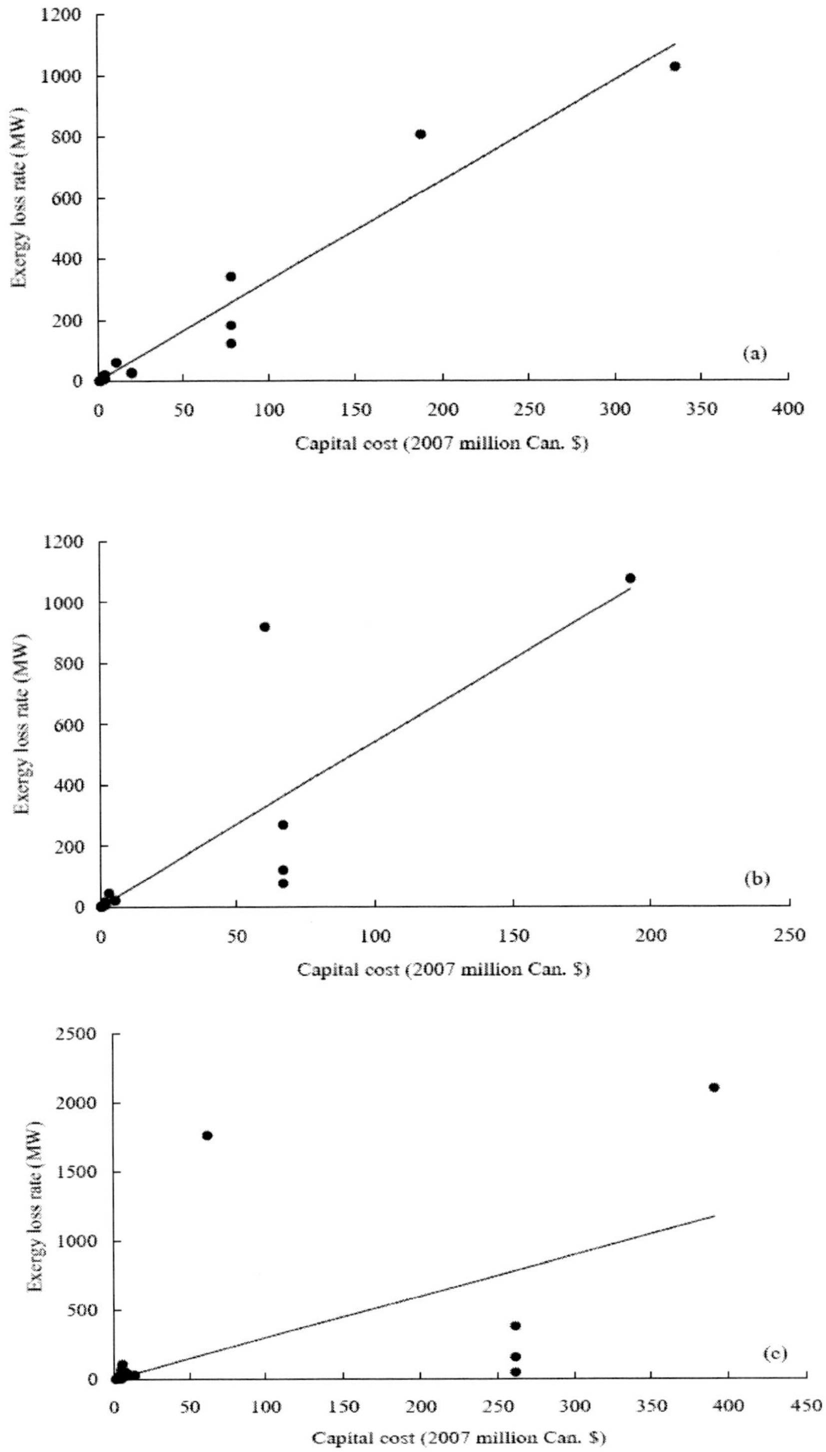

Figure 14. Exergy loss rate as a function of capital cost for several devices and the overall plant for a) coal-fired power generation, b) oil-fired power generation and c) nuclear power generation. A trend line determined using linear regression and intersecting the origin is shown to help illustrate the degree to which data points are clustered around a common slope.

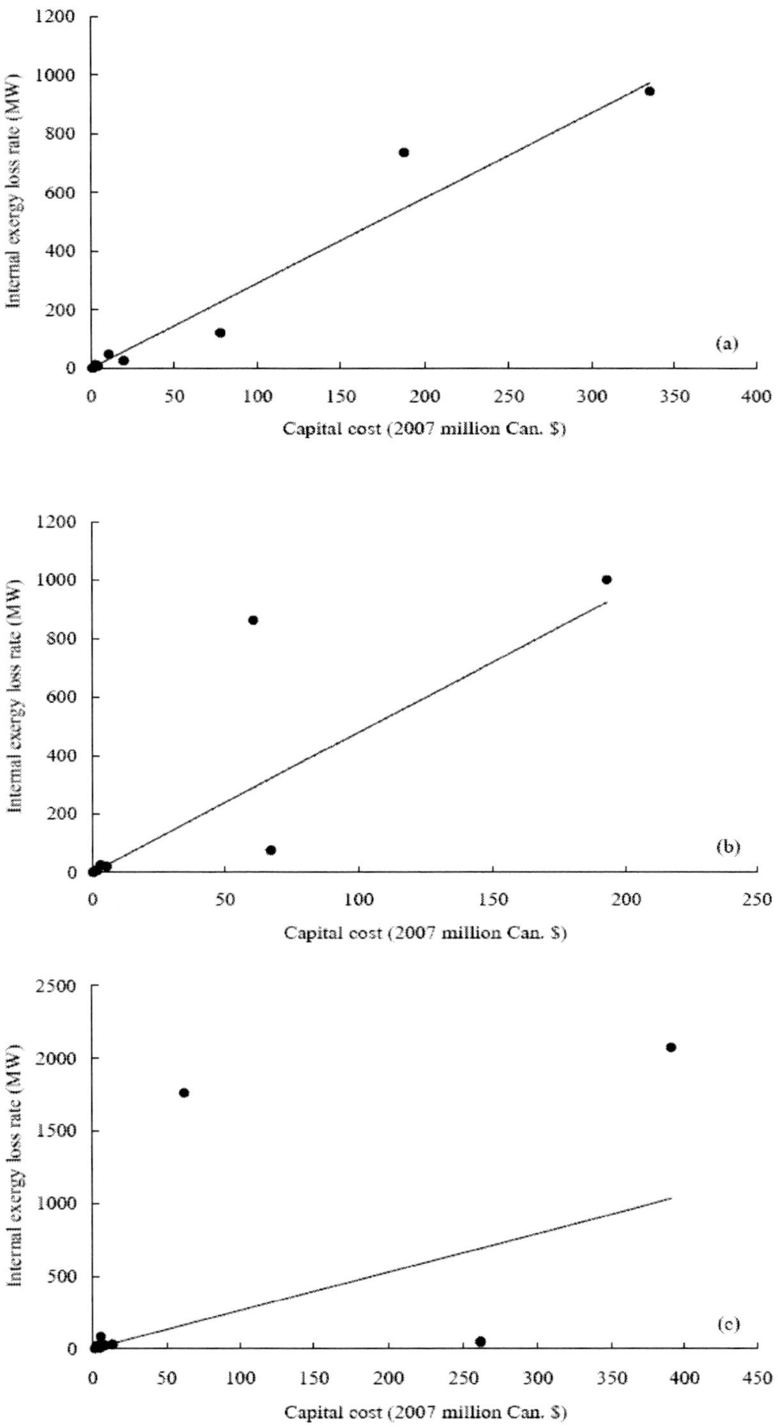

Figure 15. Internal exergy loss rate as a function of capital cost for several devices and the overall plant for a) coal-fired power generation, b) oil-fired power generation and c) nuclear power generation. A trend line determined using linear regression and intersecting the origin is shown to help illustrate the degree to which data points are clustered around a common slope.

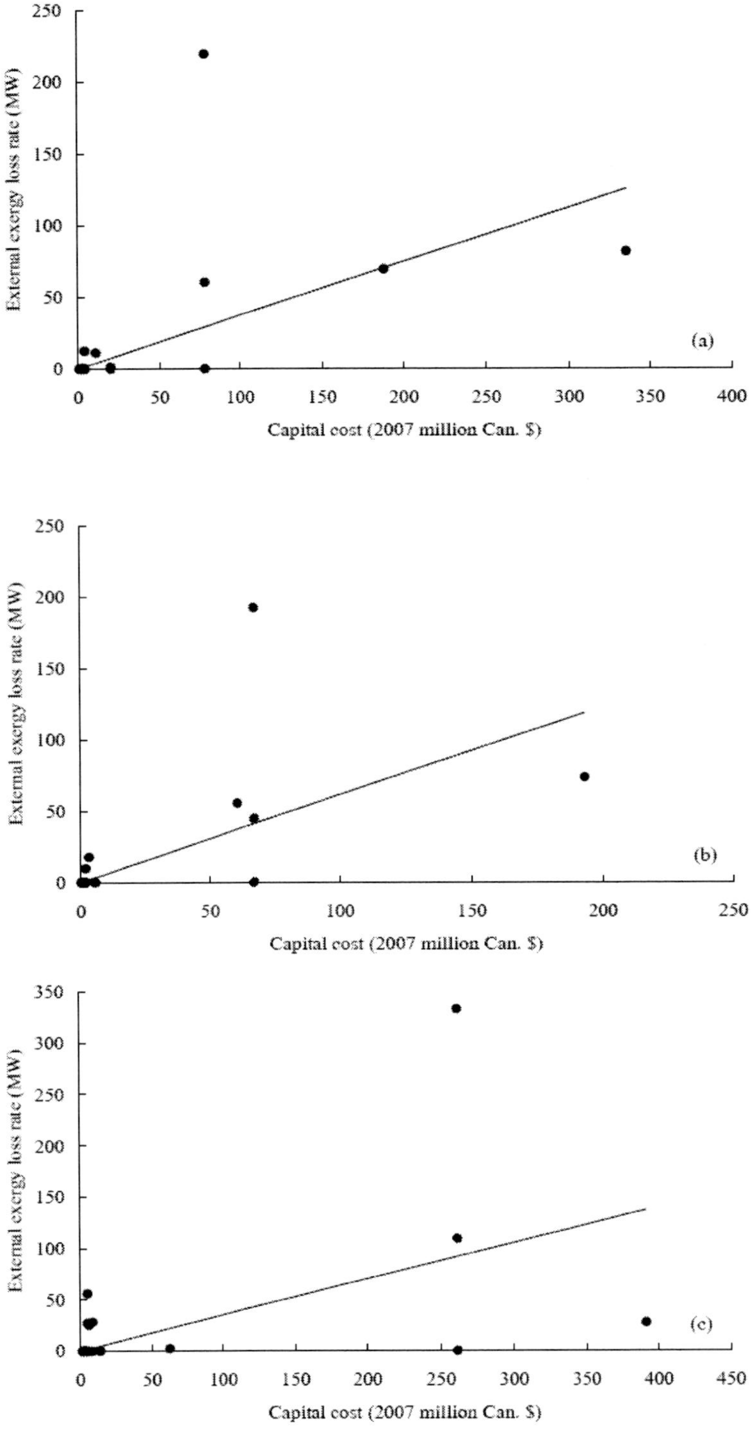

Figure 16. External exergy loss rate as a function of capital cost for several devices and the overall plant for a) coal-fired power generation, b) oil-fired power generation and c) nuclear power generation. A trend line determined using linear regression and intersecting the origin is shown to help illustrate the degree to which data points are clustered around a common slope.

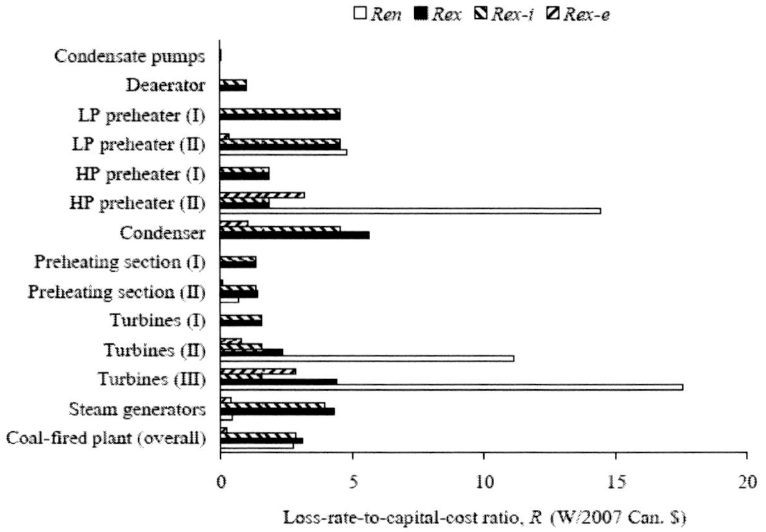

Figure 17. Values of four thermodynamic-loss-rate-to-capital-cost ratios for devices in a large coal-fired power plant and the overall plant. The bar for the condenser for *R* based on energy loss is not shown because it is far off the scale of the figure (i.e., R_{en} = 78.870 W/2007 Can. $). LP and HP denote low- and high-pressure, respectively.

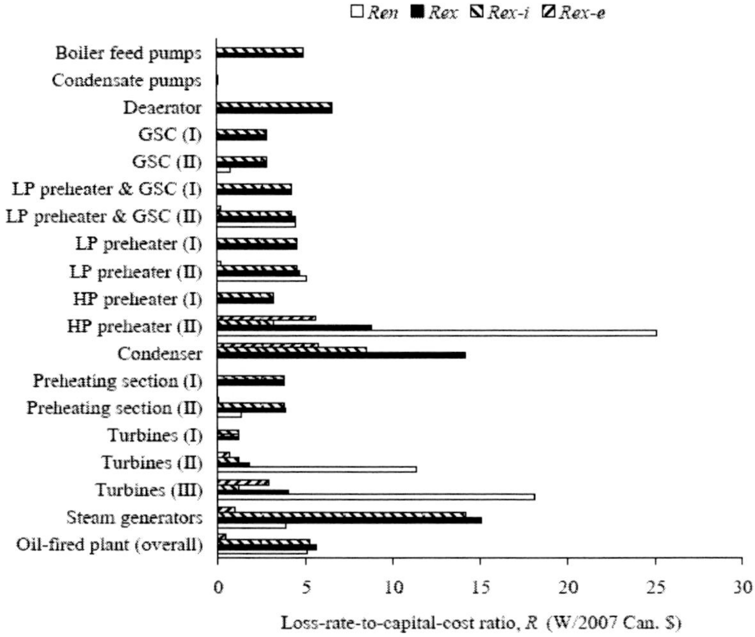

Figure 18. Values of four thermodynamic-loss-rate-to-capital-cost ratios for devices in a large oil-fired power plant and the overall plant. The bar for the condenser for *R* based on energy loss is not shown because it is far off the scale of the figure (i.e., R_{en} = 235.480 W/2007 Can. $). LP and HP denote low- and high-pressure, respectively, and GSC denotes gland seal condenser.

6.3.3. Discussion

Values for R_{ex} and $R_{ex\text{-}i}$ are seen to be similar for most cases, as are the statistical quantities calculated for sets of R_{ex} and $R_{ex\text{-}i}$ values (see Table 7). Values for R_{en} and $R_{ex\text{-}e}$ are seen to vary greatly from each other and from the values for R_{ex} and $R_{ex\text{-}i}$. The characteristics of the ratios for each type of thermal power generation considered are similar.

The statistical data in Table 7 indicate that the relative spread in thermodynamic-loss-rate-to-capital-cost ratios for different devices in each type of power plant are large when the ratio is based on energy loss (R_{en}) or external exergy loss ($R_{ex\text{-}e}$) and small when based on exergy loss (R_{ex}) or internal exergy loss ($R_{ex\text{-}i}$). This observation is supported by the range of maximum to minimum values, which shows for R_{en} and R_{ex}, for instance:

$$(R_{en})_{max} - (R_{en})_{min} >> (R_{ex})_{max} - (R_{ex})_{min} \tag{26}$$

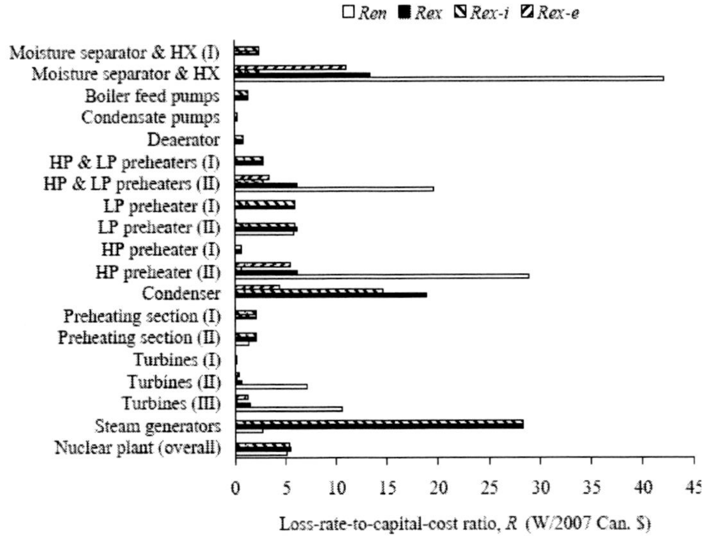

Figure 19. Values of four thermodynamic-loss-rate-to-capital-cost ratios for devices in a large nuclear power plant and the overall plant. The bar for the condenser for R based on energy loss is not shown because it is far off the scale of the figure (i.e., R_{en} = 315.392 W/2007 Can. $). LP and HP denote low- and high-pressure, respectively, and HX denotes heat exchanger.

Table 7. Data for values of loss-to-cost ratio *R* for several types of thermal power generation (using 2007 Can. $)

	Loss basis for *R*			
	Energy loss	Exergy loss		
		Total	Internal	External
Coal-fired (Nanticoke)				
Minimum, R_{min} (W/$)	0.00	0.04	0.04	0.00
Maximum, R_{max} (W/$)	78.87	5.62	4.54	3.18
Mean, R_m (W/$)	9.83	2.66	2.26	0.66
Standard deviation, SD(*R*) (W/$)	20.89	1.69	1.48	1.06

Table 7. Continued

	Loss basis for *R*			
	Energy loss	Exergy loss		
		Total	Internal	External
Coefficient of variation, CV(*R*) (%)	212.43	63.60	65.45	160.36
Oil-fired (Lennox)				
Minimum, R_{min} (W/$)	0.00	0.02	0.02	0.00
Maximum, R_{max} (W/$)	235.48	15.09	14.18	5.72
Mean, R_m (W/$)	16.33	5.07	4.20	0.87
Standard deviation, SD(*R*) (W/$)	53.52	3.88	3.13	1.81
Coefficient of variation, CV(*R*) (%)	327.64	76.46	74.67	208.39
Nuclear (Bruce B)				
Minimum, R_{min} (W/$)	0.00	0.18	0.18	0.00
Maximum, R_{max} (W/$)	315.39	28.28	28.25	10.96
Mean, R_m (W/$)	23.08	5.52	4.15	1.38
Standard deviation, SD(*R*) (W/$)	71.71	7.31	6.75	2.85
Coefficient of variation, CV(*R*) (%)	310.71	132.37	162.81	206.93
All stations combined				
Minimum, R_{min} (W/$)	0.00	0.02	0.02	0.00
Maximum, R_{max} (W/$)	315.39	28.28	28.25	10.96
Mean, R_m (W/$)	16.91	4.59	3.67	0.99
Standard deviation, SD(*R*) (W/$)	54.47	5.13	4.56	2.10
Coefficient of variation, CV(*R*) (%)	322.08	111.63	124.33	211.96

Similarly, the coefficients of variation, which indicate the relative variation, are much greater in the set of R_{en} and $R_{ex\text{-}e}$ values than in the set of R_{ex} and $R_{ex\text{-}i}$ values. For example,

$$CV(R_{en}) >> CV(R_{ex}) \tag{27}$$

For the turbines, preheating and other devices having multiple cases for wastes, the energy loss rate L_{en} and external exergy loss rate $L_{ex\text{-}e}$ are sensitive to the case. The exergy loss rate L_{ex} and internal exergy loss rate $L_{ex\text{-}i}$ are relatively less sensitive. Since the capital cost K is constant for each device for all cases considered, the corresponding thermodynamic-loss-rate-to-capital-cost ratios exhibit sensitivities similar to those exhibited by the thermodynamic loss rates.

Several important and interesting inferences for coal-fired, oil-fired and nuclear power plants can be drawn:

- A systematic correlation is evident between capital cost and exergy loss (total or internal), but not between capital cost and energy loss or external exergy loss, for devices.
- Values of R_{ex} and $R_{ex\text{-}i}$ for a group of devices and for a given type of power generation approximately conform to particular appropriate values, R^*_{ex} and $R^*_{ex\text{-}i}$. The similar behavior between R_{ex} and $R_{ex\text{-}i}$ suggests that values for R^*_{ex} and $R^*_{ex\text{-}i}$ are of similar magnitude and behave similarly. The terms R^*_{ex} and $R^*_{ex\text{-}i}$ represent appropriate optimums, in the sense of being appropriate trade-offs between exergy

losses and capital costs as practised in successful coal-fired, oil-fired and nuclear power plants. This idea is premised on the fact that such plants have been for some time widely used and the designs of an overall plant and its devices can be viewed as "successful." Certainly, the values of R^*_{ex} and R^*_{ex-i} do not identify ideal values of R_{ex} and R_{ex-i}.

- Mean thermodynamic-loss-rate-to-capital-cost ratios based on total or internal exergy loss are similar to the plant ratio values based on total or internal exergy loss. That is, $R_{m,ex} \approx (R_{ex})_s$ and $R_{m,ex-i} \approx (R_{ex-i})_s$, where the mean is for all station devices considered, and the subscript s refers to the overall station. Corresponding similarities based on either energy or external exergy are not evident. In other words, the relations between capital cost and total and internal exergy loss suggest that the collective characteristics of a plant match and benefit the overall plant. That is, devices in a successful plant appear to be arranged so as to achieve an "optimal" overall configuration, but only based on relations for capital cost and exergy loss (total and internal).

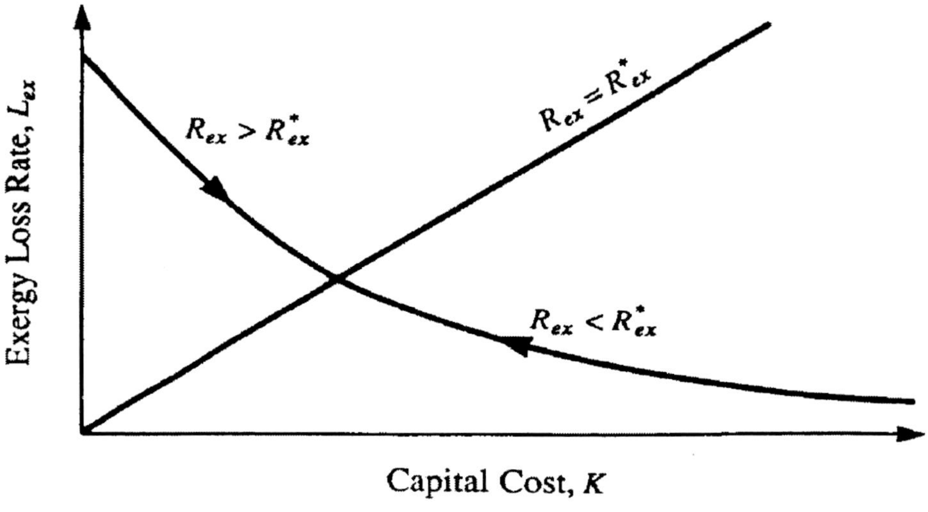

Figure 20. Plot of exergy loss rate vs. capital cost, showing the tendency of the loss-to-cost ratio R_{ex} for a device to approach R^*_{ex}, the value of R_{ex} for which an "appropriate" trade-off between losses and capital costs is attained. The curve represents the possible combinations of exergy loss rate and cost, and the intersection shows the values of capital cost and exergy loss rate at which the $R_{ex} = R^*_{ex}$.

The design of a general technology may be more successful if modified so that R_{ex} approaches R^*_{ex} for the technology. This balance between exergy loss and capital cost for a device in a real system is illustrated in Figure 20. where

- a line intersecting the origin represents R^*_{ex},
- a curve represents the continuous relation between exergy loss and capital cost for the device, and

- the intersection of the two curves, i.e., $R_{ex} = R^*_{ex}$, represents the expected combination of exergy loss and capital cost. If the cost is too low, then $R_{ex} > R^*_{ex}$, and cost should likely be increased to reduce loss, and vice versa.

The shape and position of the curve illustrates the trade-off between cost and efficiency, demonstrating that losses can normally be reduced by increasing capital investment. Two limiting cases outlined by this curve are instructive:

- The exergy input is entirely wasted if no investment is made. In other words, total exergy loss rate approaches the total exergy input rate as capital cost approaches zero.
- Efficiency approaches ideality as investment increases to very large levels. That is, the total exergy loss rate approaches zero as the capital cost approaches infinity.

The ideas in the last three paragraphs can be extended to internal exergy loss. The ideas suggest two important outcomes:

- Technologies may fail in the marketplace if they deviate too far from the appropriate R^*_{ex} to which successful technologies conform.
- Research would be potentially highly beneficial if focused on identifying devices for which the difference between the values of R_{ex} and R^*_{ex} are large and developing ways to narrow the difference.

The inference of the above observations that devices should be optimized in concert with the overall system as opposed to in "isolation," using combined economic and exergy-based methods, is not new. Others have pointed out the need to use exergy-based economics to optimise energy systems based on the relations of devices to the overall system (Hua et al., 1989).

6.3.4. Possible Extensions

The results of this case study can likely be extended in several aspects. These are discussed in this section to illustrate additional benefits that loss-cost ratio analysis may provide.

Thermal power generation sector

When data for the modes of thermal power generation considered are grouped, it is observed that $R_{m,ex} \approx (R_{m,ex})_s$ and $R_{m,ex\text{-}i} \approx (R_{m,ex\text{-}i})_s$, where $(R_{m,ex})_s$ and $(R_{m,ex\text{-}i})_s$ are the means for the overall plants (see bottom of Table 7). These correlations parallel those for the separate modes of power generation.

Consequently, the modes of power generation that make up a larger system like the electrical utility sector may behave similar to the typical plant in the sector, again based on the relations between capital costs and total or internal exergy losses, but not energy losses or internal exergy losses.

Other technologies

The relations $R_{m,ex} \approx (R_{ex})_s$ and $R_{m,ex\text{-}i} \approx (R_{ex\text{-}i})_s$ identified here may be significant characteristics not only for successful power generation, but also for other successful technologies. That is, the relations may be generalisable. Of course, different technologies must be examined before such generalizations can be conclusively drawn. This suggestion is in part based on the results of analyses of capital costs and thermodynamic losses for other devices (e.g., El-Sayed et al., 1983; Tsatsaronis and Park, 2002),

Although useful for retrofit and improvements of existing technologies, this concept would be particularly useful during design efforts for new technologies, in that the concept would provide important guidance for designers regarding the best way to introduce of new technologies.

Spatial and temporal variations

Although values for devices in a given plant of R^*_{ex} and $R^*_{ex\text{-}i}$ are likely similar to values for $R_{m,ex}$, $R_{m,ex\text{-}i}$, $(R_{ex})_s$ and $(R_{ex\text{-}i})_s$, appropriate values for these terms likely vary spatially and temporally. This observation is based on the fact that we consider power generation from different fuels using plants in different locations and built in different times here, and variations can be seen in the values $(R_{ex})_s$ and $(R_{ex\text{-}i})_s$ and in the values of $R_{m,ex}$ and $R_{m,ex\text{-}i}$.

The value of R^*_{ex} may vary for different situations because of variations in factors such as resource scarcity and costs, knowledge, ambient conditions and environmental constraints. During periods like the 1970s and 2000s when costs of energy resources increase, for example, greater capital is normally invested to reduce losses so the value of R^*_{ex} likely decreases.

Marginal costs

The results can likely be extended to marginal costs, where the marginal cost is the increase in cost resulting from saving one unit of energy or exergy (i.e., from reducing the energy or exergy loss by one unit).

Such an approach would likely indicate for devices that marginal costs based on exergy have similar values, while marginal costs based on energy vary widely.

6.4. Exergy-Based Pricing

Wall (1997) performed a case study using exergy-pricing for Göteborg, Sweden. In the case study, several common fuels and energy forms were considered. These are described below, following the energy categories introduced earlier:

- *Electricity*: The energy price is the exergy price, because electrical energy is entirely convertible to work in theory.
- *Hydrocarbon fuels*: Several hydrocarbon fuels are considered in the case study. First, several liquid fuels are considered. For gasoline, which consists mainly of octane, the exergy content is roughly 94% of the energy value. For diesel fuel, the exergy content is 42.7 MJ/kg, which is about 104% of its energy content, based on lower heating value. For fuel oil, the exergy content is approximately 97% of its energy content, based on lower heating value. For heavy oils, the exergy content is similar to

that for fuel oil. Next gaseous fuels are examined. For town gas, which is comprised of about 65% hydrogen and 20% carbon dioxide, the specific energy content is approximately 92 MJ/kg and the specific exergy content is approximately 76 MJ/kg. Thus, the exergy content of town gas is 82% of its energy content. Finally, solid hydrocarbons are considered. For coal, the exergy content and price of coal varies for each coal type. For wood and wood products, the fuel value varies significantly with water content. For wood with 50% water content, the respective energy and exergy contents are 12.4 MJ/kg and 12.1 MJ/kg (based on higher heating value, i.e., assuming the resulting water vapor condenses). Nevertheless, it is difficult to evaluate appropriate prices due to the relatively small amounts of wood and wood products sold for space heating.

- *Heat*: Heat utilized for district heating is considered in the case study.

The relative energy and exergy prices for various energy commodities studied by Wall (1997) are listed in Table 8, after converting values from SEK to US$ using the exchange rate for the year of 2000. In Table 8, the energy price for electricity is fixed at 1 $/GJ and the energy and exergy prices calculated for other commodities are adjusted relatively. Several observations are worth noting:

- The highest energy price is observed for gasoline, which is a refined fuel with special areas of use.
- Low prices are found for coal and wood products, with the lowest energy price for paper. Paper is possibly the least efficient commodity to handle as fuel and the most expensive fuel.
- The differences in energy and exergy prices in Table 8 are small for each of the energy forms except district heating.

The large difference between energy and exergy prices for district heating can be explained by noting that the exergy of an amount of district heat Q can be determined as follows:

$$Ex = Q(1 - T_{od}) / ([T_{su} - T_{od}] \ln[T_{su}/T_{re}]) \tag{28}$$

where T_{od}, T_{su} and T_{re} denote outdoor, supply and return temperatures, respectively. Clearly, the exergy content varies with the outdoor temperature. District-heating supply and return temperatures are regulated relative to the outdoor temperature. For example, Wall (1997) notes that the supply temperature is maintained at about 85°C at outdoor temperatures above 2°C and is subsequently raised in inverse proportion to the outdoor temperature, up to 120°C at an outdoor temperature of –20°C. For Göteborg, the mean exergy required for the heating season is about one-sixth of the total heat required, causing the price of exergy to be about six times more than that of energy. District heating customers in Göteborg are charged a fixed rate plus an energy rate and they pay significantly more for exergy than other energy users.

Table 8. Average energy and exergy prices (in 2000 US $) of some common energy forms for Göteborg, Sweden, relative to electricity set to 1 $/GJ

	Relative unit price ($/GJ)	
	Energy-based price	Exergy-based price
Fossil fuels and products		
Gasoline	1.67	1.78
Diesel	0.85	0.81
Fuel oil #1	0.81	0.83
Fuel oil #3-4	0.65	0.66
Coal	0.28	0.29
Town gas	0.90	1.10
Wood and products		
Firewood	0.29	0.29
Paper	0.21	0.22
Wood paper	0.47	0.49
Intermediate energy commodities		
Electricity	1.00	1.00
District heat	0.83	4.89

Adapted from Wall (1997).

Means exist to lower energy prices. For instance, cogeneration of electricity and heat can lead to reduced overall costs through increased efficiencies. Also, the use of a heat pump can lower heating bills, as heat pumps normally permit efficient and cost-effective heat generation. Heat pumps can be employed at user locations instead of district heating, or as part of district heating systems. With heat pumps, the energy and exergy prices for district heating in Table 8 would change.

The case study suggests that appropriate energy policies could accelerate movement towards efficient and resource-saving technologies by basing prices on exergy, rather than energy.

7. Conclusion

Useful relations exist between exergy and economics. These relations are different from the relations between energy and economics in important ways, and provide the basis for exergy-based economic analysis methods. Such methods include exergoeconomics and thermoeconomics, exergy-based pricing, EXCEM analysis and analysis based on the ratio of thermodynamic loss to capital cost, and are useful in analysis and improvement activities. The methods generally combine thermodynamics with economics to achieve advantageous designs, accounting for observations that costs are better distributed among outputs if cost accounting is based on exergy rather than energy, and a consistent measure of economic value is provided by exergy but not energy. Loss-cost ratio analysis is based on the important correlation between the ratio of thermodynamic loss rate to capital cost when total or internal exergy losses are considered. This correlation appears to imply that devices in successful systems are configured so as to achieve an overall optimal design by appropriately balancing exergy-based thermodynamic and economic characteristics of the overall system and its devices. Generally, exergy-based economic analysis methods usually determine appropriate economic-resource allocations for the design and operation of a system, and economic feasibility and profitability.

This chapter also highlights the merits of exergy analysis over the more conventional energy analysis, from a thermodynamic perspective and also from a combined thermodynamic and economic perspective. Proponents of exergy analysis and exergy-based economic methods conventionally argue that they can help improve process performance. For existing "mature" technologies which have been gradually improved over time, sometimes in a non-systematic way, however, the application of exergy analysis and exergy-based economic methods may not lead to significant design modifications or performance improvements. Consequently, these demonstrations alone do not convince many non-users of exergy analysis and exergy-based economic methods of their merits. But for new technologies, these methods are likely to assist in improving designs quickly, making it important that the merits of exergy analysis and exergy-based economic methods be appreciated.

References

Ahrendts, J. (1980). Reference states. *Energy-The International Journal*, 5, 667-678.

Ayres, R. U. (1998). Eco-thermodynamics: Economics and the second law. *Ecological Economics*, 26, 189-209.

Ayres, R. U., Ayres, L. W., and Warr, B. (2003). Exergy, power and work in the US economy, 1900-1998. *Energy-The International Journal*, 28, 219-273.

Baehr, H. D., and Schmidt, E. F. (1963). Definition und berechnung von brennstoffexergien (Definition and calculation of fuel exergy). *Brennst-Waerme-Kraft*, 15, 375-381.

Bakan, K., Dincer, I., and Rosen, M. A. (2008). Exergoeconomic analysis of glycol cold thermal energy storage systems. *Int. J. Energy Research*, 32, 215-225.

Bandura, A. V., and Brodiansky, V. M. (2001). Thermodynamics extends economics potentials. *Energy*, 26, 811-814.

Bejan, A. (1982). *Entropy Generation through Heat and Fluid Flow*. Toronto: Wiley.

Bejan, A. (2006). *Advanced Engineering Thermodynamics* (3rd ed.). New York: John Wiley and Sons.

Bejan, A., Tsatsaronis, G., and Moran, M. (1996). *Thermal Design and Optimization*. New York: John Wiley and Sons.

Borgert, J. A., and Velasquez, J. A. (2004). Exergoeconomic optimisation of a Kalina cycle for power generation. *International Journal of Exergy*, 1, 18-28.

Bosnjakovic, F. (1963). Bezugszustand der exergie eines reagiernden systems (Reference states of the exergy in a reacting system). *Forsch. Ingenieurw*, 20, 151-152.

Boyarskii, M. Y., Garin, V. A., Smorodin, A. I., and Matveev, S. A. (1992). Thermal and economic analysis of costs in obtaining ultrapure products from air separators. *Chemical and Petroleum Engineering*, 28, 497-500.

Brodyanski, V. M., Sorin, M. V., and Le Goff, P. (1994). *The Efficiency of Industrial Processes: Exergy Analysis and Optimization*. London: Elsevier.

Bryant, J. (2007). A thermodynamic theory of economics. *International Journal of Exergy*, 4, 302-337.

Colpan, C. O., and Yesin, T. (2006). Thermodynamic and thermoeconomic comparison of combined cycle cogeneration systems. *International Journal of Exergy*, 3, 272-290.

Connelly, L., and Koshland, C. P. (2001a). Exergy and industrial ecology. Part 1: An exergy-based definition of consumption and a thermodynamic interpretation of ecosystem evolution. *Exergy, An International Journal*, 3, 146-165.

Connelly, L., and Koshland, C. P. (2001b). Exergy and industrial ecology. Part 2: A non-dimensional analysis of means to reduce resource depletion. *Exergy, An International Journal*, 1, 234-255.

Cownden, R., Nahon, M., and Rosen, M. A. (2001). Exergy analysis of a fuel cell power system for transportation applications. *Exergy, An International Journal*, 1, 112-121.

Dewulf, J. P., and Van Langenhove, H. R. (2002). Quantitative assessment of solid waste treatment systems in the industrial ecology perspective by exergy analysis. *Environ. Sci. Technol.*, 36, 1130-1135.

Dincer, I. (2002). On energetic, exergetic and environmental aspects of drying systems. *Int. J. Energy Research*, 26, 717-727.

Dincer, I., Hussain, M. M., and Al-Zaharnah, I. (2003). Energy and exergy use in industrial sector of Saudi Arabia. *Proceedings of the Institution of Mechanical Engineers, Part A: Journal of Power and Energy*, 217, 481-492.

Dincer, I., and Rosen, M. A. (2007). *Exergy: Energy, Environment and Sustainable Development*. Oxford, UK: Elsevier.

Dipippo, R. (1994). Second law analysis of flash-binary and multilevel binary geothermal plants. *Geothermal Resources Council Transactions*, 18, 505-510.

Edgerton R. H. (1982). *Available Energy and Environmental Economics*. Toronto: D. C. Heath.

El-Sayed, Y. (2004). *The Thermoeconomics of Energy Conversion*. Elsevier: Amsterdam.

El-Sayed, Y. M., and Tribus, M. (1983). Strategic use of thermoeconomics for system improvement. In R. A. Gaggioli (Ed.), *Efficiency and Costing: Second Law Analysis of Processes* (*ACS Symposium Series* 235, pp. 215-238). Washington, DC: American Chemical Society.

Erlach, B., Serra, L., and Valero, A. (1999). Structural theory as standard for thermoeconomics. *Energy Conversion and Management*, 40, 1627-1649.

Ferrari, S., Genoud, S., and Lesourd, J.-B. (2001). Thermodynamics and economics: Towards exergy-based indicators of sustainable development. *Schweizerische Zeitschrift für Volkswirtschaft und Statistik*, 137, 319-336.

Gaggioli, R. A. (1983). Second law analysis to improve process and energy engineering. In *Efficiency and Costing: Second Law Analysis of Processes* (*ACS Symposium Series* 235, pp. 3-50). Washington, DC: American Chemical Society.

Gaggioli, R. A. (1998). Available energy and exergy. *International Journal of Applied Thermodynamics*, 1, 1-8.

Gaggioli, R. A., and Petit, P. J. (1977). Use the second law first. *Chemtech*, 7, 496-506.

Georgescu-Roegen, N. (1971). *The Entropy Law and Economic Process*. Cambridge: Harvard University Press.

Gogus, Y. A. (2005). Thermoeconomic optimization. *International Journal of Energy Research*, 29, 559-580.

Granovskii, M., Dincer, I., and Rosen, M. A. (2006). Economic and environmental comparison of conventional, hybrid, electric and hydrogen fuel cell vehicles. *J. Power Sources*, 159, 1186-1193.

Gunnewiek, L. H., and Rosen, M. A. (1998). Relation between the exergy of waste emissions and measures of environmental impact. *International Journal of Environment and Pollution*, 10, 261-272.

Hammond, G. P., and Stapleton, A. J. (2001). Exergy analysis of the United Kingdom energy system. *Proceedings of the Institution of Mechanical Engineers, Part A: Journal of Power and Energy*, 215, 141-162.

Honkasalo, A. (1998). Entropy, exergy and steady-state economy. *Sustainable Development*, 6, 130-142.

Hua, B., Yin, Q., and Wu, G. (1989). Energy optimization through exergy-economic evaluation. *ASME J. Energy Resources Technology*, 111, 148-153.

Jaber, M. Y., Nuwayhid, R. Y., and Rosen, M. A. (2004). Price-driven economic order systems from a thermodynamic point of view. *Int. J. Production Research*, 42, 5167-5184.

Jaber, M. Y., Nuwayhid, R. Y., and Rosen, M. A. (2006). A thermodynamic approach to modelling the economic order quantity. *Applied Mathematical Modelling*, 30, 867-883.

Jaber, M. Y., and Rosen, M. A. (2008). The economic order quantity repair and waste disposal model with entropy cost. *European Journal of Operational Research*, 188, 109-120.

Jassim, R. K., and Khir, T. (2004). Exergoeconomic optimisation of an air-conditioning rotary regenerator: Effect of matrix thermal conductivity on its performance. *International Journal of Exergy*, 1, 215-236.

Jassim, R. K., Khir, T., and Ghaffour, N. (2005). Thermoeconomic optimization of the geometry of an air conditioning precooling air reheater dehumidifier. *International Journal of Energy Research*, 30, 237-258.

Jassim, R. K., Khir, T., Habeebullah, B. A., and Zaki, G. M. (2005). Exergoeconomic optimisation of the geometry of continuous fins on an array of tubes of a refrigeration air cooled condenser. *International Journal of Exergy*, 2, 146-171.

Jorgensen, S. E. (2000). Application of exergy and specific exergy as ecological indicators of coastal areas. *Aquatic Ecosystem Health and Management*, 3, 419-430.

Jorgensen, S. E., and Fath, B. D. (2004). Application of thermodynamic principles in ecology. *Ecological Complexity*, 1, 267-280.

Kanoglu, M., Carpinlioglu, M. O., and Yildirim, M. (2004). Energy and exergy analyses of an experimental open-cycle desiccant cooling system. *Applied Thermal Engineering*, 24, 919-932.

Kay, J. (2002). On complexity theory, exergy, and industrial ecology: Some implications for construction ecology. In C. Kibert, J. Sendzimir, and B. Guy (Eds.), *Construction Ecology: Nature as a Basis for Green Buildings* (pp. 72-107). London: Spon Press.

Kazim, A. (2005). Exergoeconomic analysis of a PEM fuel cell at various operating conditions. *Energy Conversion and Management*, 46, 1073-1081.

Kestin, J. (1980). Availability: The concept and associated terminology. *Energy-The International Journal*, 5, 679-692.

Koroneos, C., Spachos, N., and Moussiopoulos, N. (2003). Exergy analysis of renewable energy sources. *Renewable Energy*, 28, 295-310.

Kotas, T. J. (1995). *The Exergy Method of Thermal Plant Analysis* (reprint ed.). Malabar, Florida: Krieger.

Lazzaretto, A., and Toffolo, A. (2004). Energy, economy and environment as objectives in multi-criterion optimization of thermal systems design. *Energy*, 29, 1139-1157.

Lazzaretto, A., and Tsatsaronis, G. (2006). SPECO: A systematic and general methodology for calculating efficiencies and costs in thermal systems. *Energy*, 31, 1257-1289.

Lee, K. C. (2001). Classification of geothermal resources by exergy. *Geothermics*, 30, 431-442

Luz-Silveira, J., Beyene, A., Leal, E. M., Santana, J. A., and Okada, D. (2002). Thermoeconomic analysis of a cogeneration system of a university campus. *Applied Thermal Engineering*, 22, 1471-1483.

Mazur, V. A. (2005). Fuzzy thermoeconomic optimization. *International Journal of Exergy*, 2, 1-13.

Moran, M. J. (1989). *Availability Analysis: A Guide to Efficient Energy Use* (revised ed.). New York: American Society of Mechanical Engineers.

Moran, M. J. and Shapiro, H. N. (2007). *Fundamentals of Engineering Thermodynamics* (6th ed.). New York: Wiley.

Nafey, A. S., Fath, H. E. S., and Mabrouk, A. A. (2006). Thermo-economic investigation of multi effect evaporation (MEE) and hybrid multi effect evaporation-multi stage flash (MEE-MSF) systems. *Desalination*, 201, 241-254.

Nuwayhid, R. Y., Jaber, M. Y., Rosen, M. A., and Sassine, G. P. (2006). On the thermodynamic treatment of diffusion-like economic commodity flows. *International Journal of Exergy*, 3, 103-117.

Ontario Hydro. (1969). Internal and External Manhour and Cost Comparisons for Fossil and Nuclear Projects. Report.

Ontario Hydro. (1973). Nanticoke, data sheet no. PRD2387-73-10m.

Ontario Hydro. (1979). Reports Lennox Generating Station. Lakeview Generating Station Units 7 and 8 Completion Costs.

Ontario Hydro. (1983). Nanticoke Generating Station, data sheet.

Ontario Hydro. (1985). Pickering Generation Station Data: Units 1–4, data sheet no. 966 A119:0.

Ontario Hydro. (1991). Final Costs. Information document.

Ontario Hydro. (1996). Nanticoke Generating Station Technical Data. Information document.

Ozdogan, O., and Arikol, M. (1995). Energy and exergy analyses of selected Turkish industries. *Energy-The International Journal*, 20, 73-80.

Ozgener, O., and Hepbasli, A. (2005). Exergoeconomic analysis of a solar assisted ground-source heat pump greenhouse heating system. *Applied Thermal Engineering*, 25, 1459-1471.

Ozgener, L., Hepbasli, A., and Dincer, I. (2004). Thermo-mechanical exergy analysis of Balcova geothermal district heating system in Izmir, Turkey. *Transactions of the ASME, Journal of Energy Resources Technology,* 126, 293-301.

Ozgener, O., Hepbasli, A., Dincer, I., and Rosen, M. A. (2005). Modelling and assessment of ground-source heat pump systems using exergoeconomic analysis for building applications. *Proc. Building Simulation 2005: 9th International Building Performance Simulation Association Conf.*, 15-18 Aug., Montreal, pp. 915-920.

Ozgener, L., Hepbasli, A., Dincer, I., and Rosen, M. A. (2007). Exergoeconomic analysis of geothermal district heating systems: A case study. *Applied Thermal Engineering*, 2, 1303-1310.

Ozturk, I. T., Karabay, H., and Bilgen, E. (2006). Thermo-economic optimization of hot water piping systems: A comparison study. *Energy*, 31, 2094-2107.

Reistad, G. M. (1975). Available energy conversion and utilization in the United States. *J. Eng. Power*, 97, 429-434.

Reistad, G. M., and Gaggioli, R. A. (1980). Available energy costing. In *Thermodynamics: Second Law Analysis* (*ACS Symposium Series* 122, pp. 143-159). Washington, DC: American Chemical Society.

Rensonnet, T., Uche, J., and Serra, L. (2007). Simulation and thermoeconomic analysis of different configurations of gas turbine (GT)-based dual-purpose power and desalination plants (DPPDP) and hybrid plants (HP). *Energy*, 32, 1012-1023.

Rezac, P., and Metghalchi, H. (2004). A brief note on the historical evolution and present state of exergy analysis. *Int. J. Exergy*, 1, 426-437.

Rodriguez, L. S. J. (1980). Calculation of available-energy quantities. In *Thermodynamics: Second Law Analysis* (*ACS Symposium Series* 122, 39–60). Washington, DC: American Chemical Society.

Rosen, M. A. (1990). The relation between thermodynamic losses and capital costs for a modern coal-fired electrical generation station. In *Computer-Aided Energy Systems Analysis* (AES-Vol. 21, pp. 69-78). New York: American Society of Mechanical Engineers.

Rosen, M. A. (1991). An investigation of the relation between capital costs and selected thermodynamic losses. In *Second Law Analysis: Industrial and Environmental Applications* (AES-Vol. 25/HTD-Vol. 191, pp. 55-62). New York: American Society of Mechanical Engineers.

Rosen, M. A. (1992). Evaluation of energy utilization efficiency in Canada using energy and exergy analyses. *Energy-The International Journal*, 17, 339-350.

Rosen, M. A. (2001). Energy- and exergy-based comparison of coal-fired and nuclear steam power plants. *Exergy, An International Journal*, 1, 180-192.

Rosen, M. A. (2002a). Can exergy help us understand and address environmental concerns? *Exergy, An International Journal*, 2, 214-217.

Rosen, M.A. (2002b). Exergy and economics: Is exergy profitable? *Exergy, An International Journal*, 2, 218-220.

Rosen, M. A., and Dincer, I. (1997a). On exergy and environmental impact. *International Journal of Energy Research*, 21, 643-654.

Rosen, M. A., and Dincer, I. (1997b). Sectoral energy and exergy modeling of Turkey. *ASME J. Energy Resources Technology*, 119, 200-204.

Rosen, M. A., and Dincer, I. (1999). Exergy analysis of waste emissions. *International Journal of Energy Research*, 23, 1153-1163.

Rosen, M. A., and Dincer, I. (2002). Energy and exergy analyses of thermal energy storage systems. In *Thermal Energy Storage: Systems and Applications*, Chapter 10, pp. 411-510). London, Wiley.

Rosen, M. A., and Dincer, I. (2003a). Thermoeconomic analysis of power plants: An application to a coal-fired electrical generating station. *Energy Conversion and Management*, 44, 2743-2761.

Rosen, M. A., and Dincer, I. (2003b). Exergoeconomic analysis of power plants operating on various fuels. *Applied Thermal Engineering*, 23, 643-658.

Rosen, M. A., and Dincer, I. (2003c). Exergy-cost-energy-mass analysis of thermal systems and processes. *Energy Conversion and Management*, 44, 1633-1651.

Rosen, M. A., and Dincer, I. (2004). A study of industrial steam process heating through exergy analysis. *Int. J. Energy Research*, 28, 917-930.

Rosen, M. A., and Etele, J. (2004). Aerospace systems and exergy analysis: Applications and methodology development needs. *Int. J. Exergy*, 1, 411-425.

Rosen, M. A., and Horazak, D. A. (1995). Energy and exergy analyses of PFBC power plants. In M. Alvarez Cuenca and E. J. Anthony (Eds.), *Pressurized Fluidized Bed Combustion* (pp. 419-448). London: Chapman and Hall.

Rosen, M. A., Le, M. N., and Dincer, I. (2005). Efficiency analysis of a cogeneration and district energy system. *Applied Thermal Engineering*, 25, 147-159.

Rosen, M. A., Tang, R., and Dincer, I. (2004). Effect of stratification on energy and exergy capacities in thermal storage systems. *Int. J. Energy Research*, 28, 177-193.

Sahoo, P. K., Misra, R. D., and Gupta, A. (2004). Exergoeconomic optimisation of an aqua-ammonia absorption refrigeration system. *International Journal of Exergy*, 1, 82-93.

Santarelli, M. G. L. (2004). Carbon exergy tax: A thermo-economic method to increase the efficient use of exergy resources. *Energy Policy*, 32, 413-427.

Santarelli, M., and Macagno, S. A. (2004). A thermoeconomic analysis of a PV-hydrogen system feeding the energy requests of a residential building in an isolated valley of the Alps. *Energy Conversion and Management*, 45, 427-451.

Sato, N. (2005). *Chemical Energy and Exergy: An Introduction to Chemical Thermodynamics for Engineers*. Oxford, UK: Elsevier.

Sciubba, E. (1999). Exergy as a direct measure of environmental impact. *Proceedings of the ASME Advanced Energy Systems Division, American Society of Mechanical Engineers*, 39, 573-581.

Sciubba, E. (2001). Beyond thermoeconomics? The concept of Extended Exergy Accounting and its application to the analysis and design of thermal systems. *Exergy, An International Journal*, 1, 68-84.

Sciubba, E. (2003). Cost analysis of energy conversion systems via a novel resource-based quantifier. *Energy*, 28, 457-477.

Sciubba, E. (2004). From engineering economics to extended exergy accounting: A possible path from monetary to resource-based costing. *J. Industrial Ecology*, 8, 19-40.

Sciubba, E. (2005). Exergo-economics: Thermodynamic foundation for a more rational resource use. *International Journal of Energy Research*, 29, 613-636.

Sugimoto, K., Fujii, T., and Ohta, J. (2006). An appraisal method of exergy cost minimisation for co-generation systems. *International Journal of Exergy*, 3, 255-271.

Sussman, M. V. (1980). Steady-flow availability and the standard chemical availability. *Energy-The International Journal*, 5, 793-804.

Sussman, M. V. (1981). Second law efficiencies and reference states for exergy analysis. *Proc. 2nd World Congress Chem. Eng.*, Canadian Society of Chemical Engineers, Montreal, 420-421.

Szargut, J. (1967). Grezen fuer die anwendungsmoeglichkeiten des exergiebegriffs (Limits of the applicability of the exergy concept). *Brennst.-Waerme-Kraft*, 19, 309-313.

Szargut, J. (1980). International progress in second law analysis. *Energy*, 5, 709-718.

Szargut, J. (2005). *Exergy Method: Technical and Ecological Applications*. Southampton, UK: WIT Press.

Szargut, J., Morris, D. R, and Steward F. R. (1988). *Exergy Analysis of Thermal, Chemical, and Metallurgical Processes*. New York: Hemisphere.

Szargut, J., Ziebik, A., and Stanek, W. (2002). Depletion of the non-renewable natural resources as a measure of the ecological cost. *Energy Conversion and Management*, 43, 1149-1163.

Toffolo, A., and Lazzaretto, A. (2002). Evolutionary algorithms for multi-objective energetic and economic optimization in thermal system design. *Energy*, 27, 549-567.

Tonon, S., Brown, M. T., Luchi, F., Mirandola, A., Stoppato, A., and Ulgiati, S. (2006). An integrated assessment of energy conversion processes by means of thermodynamic, economic and environmental parameters. *Energy*, 31, 149-163.

Torres, C., Valero, A., Serra, L., and Royo, J. (2002). Structural theory and thermoeconomic diagnosis. Part I: On malfunction and dysfunction analysis. *Energy Conversion and Management*, 43, 1503-1518.

Tribus, M., and Mclrivne, E. C. (1971). Energy and information. *Sci. American*, 225, 3, 179-188.

Tsatsaronis, G. (1987). A review of exergoeconomic methodologies. In *Second Law Analysis of Thermal Systems* (pp. 81-87). New York: American Society of Mechanical Engineers.

Tsatsaronis, G. (1993). Thermoeconomic analysis and optimization of energy systems. *Progress in Energy and Combustion Science*, 19, 227-257.

Tsatsaronis, G. (1996). Exergoeconomics: Is it only a new name? *Chemical Engineering and Technology*, 19, 163-169.

Tsatsaronis, G., and Park, M. H. (2002). On avoidable and unavoidable exergy destructions and investment costs in thermal systems. *Energy Conversion and Management*, 43, 1259-1270.

Tsatsaronis, G., and Valero, A. (1989). Thermodynamics meets economics. *Mechanical Engineering*, August, 84-86.

Tsatsaronis, G., and Winhold, M. (1985). Exergoeconomic analysis and evaluation of energy conversion plants. Parts 1 and 2. *Energy-The International Journal*, 10, 69-94.

Tyagi, S. K., Wang, S. W., Chen, G. M., Wang, Q., Chandra, H., and Wu, C. (2007). Performance investigations under maximum ecological and maximum economic conditions of a complex Brayton cycle. *International Journal of Exergy*, 4, 98-116.

Uche, J., Serra, L., and Valero, A. (2006). Exergy costs and inefficiency diagnosis of a dual-purpose power and desalination plant. *Transactions of the ASME: Journal of Energy Resources Technology*, 128, 186-193.

Valero, A. (2006). Exergy accounting: Capabilities and drawbacks. *Energy*, 31, 164-180.

Valero, A., Lerch, F., Serra, L., and Royo, J. (2002). Structural theory and thermoeconomic diagnosis. Part II. Application to an actual power plant. *Energy Conversion and Management*, 43, 1519-1535.

Valero, A., Serra, L., and Uche, J. (2006a). Fundamentals of exergy cost accounting and thermoeconomics. Part I: Theory. *Transactions of the ASME: Journal of Energy Resources Technology*, 128, 1-8.

Valero, A., Serra, L., and Uche, J. (2006b). Fundamentals of exergy cost accounting and thermoeconomics. Part II: Applications. *Transactions of the ASME: Journal of Energy Resources Technology*, 128, 9-15.

Van Den Bulck, E., Klein, S. A., and Mitchell, J. W. (1988). Second law analysis of solid desiccant rotary dehumidifiers. *ASME Journal of Heat Transfer*, 110, 2-9.

Verda, V., Serra, L., and Valero, A. (2005a). Thermoeconomic diagnosis: Zooming strategy applied to highly complex energy systems. Part 1: Detection and localization of anomalies. *ASME Journal of Energy Resources Technology*, 127, 42-49.

Verda, V., Serra, L. and Valero, A. (2005b). Thermoeconomic diagnosis: Zooming strategy applied to highly complex energy systems. Part 2: On the choice of the productive structure. *ASME Journal of Energy Resources Technology*, 127, 50-58.

Wall, G. (1990). Exergy conversion in the Japanese Society. *Energy-The International Journal*, 15, 435-444.

Wall, G. (1997). Exergy use in the Swedish society 1994. In *Proceedings of the International Conference on Thermodynamic Analysis and Improvement of Energy Systems*, June 10-13, Beijing, China, pp. 403-413.

Wall, G. (2003). Exergy tools. *Proceedings of the Institution of Mechanical Engineers, Part A: Journal of Power and Energy*, 217, 125-136.

Wall, G., Sciubba, E., and Naso, V. (1994). Exergy use in the Italian society. *Energy-The International Journal*, 19, 1267-1274.

Wepfer W. J., and Gaggioli, R. A. (1980). Reference datums for available energy. In *Thermodynamics: Second Law Analysis* (*ACS Symposium Series* 122, pp. 77-92). Washington, DC: American Chemical Society.

Yantovskii, E. I. (1994). *Energy and Exergy Currents: An Introduction to Exergonomics*. New York: Nova Science Publishers.

Zhang, C., Chen, S., Zhen, C., and Lou, X. (2007). Thermoeconomic diagnosis of a coal fired power plant. *Energy Conversion and Management*, 48, 405-419.

Zhang, C., Wang, Y., Zhen, C., and Lou, X. (2006). Exergy cost analysis of a coal fired power plant based on structural theory of thermoeconomics. *Energy Conversion and Management*, 47, 817-843.

In: New Research on Energy Economics
Editors: P.G. Caldwell and E.V. Taylor, pp. 67-106
ISBN: 978-1-60456-354-2

Chapter 2

ENERGY ECONOMICS AND THE ECOLOGY OF SCALE IN THE FOOD BUSINESS

Elmar H. Schlich
Dept. of Process Engineering, Justus-Liebig-University Giessen, Giessen, Germany

Abstract

Food of local origin is often said to be environmentally friendly because of small transport distances. In public opinion, long distance transportation is equated with high energy turnover, without asking for the means of transportation. But the specific energy turnover decreases with increasing size of the transportation unit. One commuter bus requires much less energy compared with 40 cars and one HGV is more effective than 100 light trucks. Looking at ecological impacts, the energy turnover is one of the most important matters, and is closely related to greenhouse impact also.

From economics we know, that the specific financial effort depends upon the number of items which are produced, because of the fixed costs to be taken into account. This well-known economics law is called "The Economy of Scale". Regarding the ecology, we actually assume the same regularity as in economics: Above all, the specific ecological impact should be dependent on the number of produced items. We assume small business units to be less effective in terms of ecology compared with bigger units. And it should be possible to derive a minimum business size as a "break even" area. This central hypothesis we call "The Ecology of Scale".

To investigate our hypothesis in the food business, case studies with juices, lamb meat, beef, pork, wine, and apples have been performed. In all cases food of local origin is compared to global food. In the first step, the energy turnover of the entire process chain has been investigated, including farming, crop and breed, transports of raw, intermediate, and final products as well as the efforts of distribution up to the point of sale. All energy efforts are allocated to the functional units.

The results demonstrate a declining relation of the specific energy turnover and the business size. Small units require more energy per food item compared with bigger enterprises, independent of local or global origin. Surprisingly, we can even find local farmers, which can cause higher specific energy turnover compared with some global process chains, because of high sophisticated logistics and good energy saving strategies of bigger units. But – on the other hand – regional food can compete with global business in terms of energy in the case of sufficient business size. Additionally, the researched data indicate a minimum business size as break even. All these results support the theory of Ecology of

Scale. In all case studies, we find regional food business meeting these requirements as well. But, there is no indication to blame global food as energy wasting process chains.

Keywords: Energy, Food, Ecology of Scale, Process chain, Globalization, Regionalism.

Introduction

Germany as a nation belongs to the industrial nations. The German economy is characterized by strong exports of machines, engineering, and technology. At the same time, the density of population is very high. In Germany, 82.5 million people are living on 352,000 km^2, corresponding to about 231 inhabitants per km^2 on average. Regarding other industrial nations, only the Netherlands (395 inhabitants per km^2), Japan (337 inhabitants per km^2) and the United Kingdom (249 inhabitants per km^2) show a higher density of population. Further, highly industrialized countries like France (112 people per km^2), the United States (30 people per km^2) or Sweden (20 people per km^2) cannot be compared with Germany in terms of population density [StaBu 2007].

Obviously, Germany is no agrarian country. On the contrary, Germany necessarily depends on the import of food, because neither the agrarian area nor the climate conditions are sufficient for the supply of the entire population with food. By the way, these natural conditions caused some emigration in the 19th century, because natural resources were too small and people were starving at that time. Nowadays, people in Germany are well used to being supplied with all different kinds of food at all times of the year.

Food, bound or retail in Germany, is originating from four different sources: local German areas, Germany, the European Union and overseas. In the present context, we define food from these sources as local food, regional food, European-continental food, and global food. Table 1 gives an overview of those definitions and shows marketing distances.

All these different sources of food require sophisticated process chains, which consist of many single steps, in order to transport raw material from primary production to food facilities, to transport intermediate and final products to the trade posts and the retailers and – last but not least – to distribute the consumer goods to the private consumers.

Table 1. Food categories, origin and marketing distances in Germany

Food category	Origin	Marketing distance
Local food	Local areas	< 50 km
Regional food	Germany	< 500 km
European-continental food	European Union (EU)	500 – 2,500 km
Global food	Outside EU	> 2,500 km

At first sight, it seems obvious, why food of local or regional origin is often said to be environmental friendly because of rather small transport distances from the place of origin to the point of sale. At the same time, in the German public opinion, long distance transports are quite often equated with high energy turnover. Usually people believe that local or regional production and consumption of food require all over less energy than European-continental or even global process chains in the food business. Understandably, this public opinion is widely supported by political statements of local or regional German government and by promotion tours of agrarian lobbyists such as CMA[1].

From the scientific point of view there is only one case where there is no doubt about the environmentally friendly quality of a process chain: This is the proverbial "*apple from my garden*". It is clear, that digging potatoes in my own garden, and harvesting apples from my own apple-tree in the backyard, or squeezing grapes from the private vineyard behind the house does not cause any transport and distribution efforts with measurable ecological impacts. But all these examples belong to process chains characterized by manual labour, by small amounts of crop and of course by seasonal supply, only.

Machinery or even industrial manufacturing does not play any role in that hobby area. As soon as bigger amounts of crop and breed require professional farming, breeding and crop, then professional transporting and distributing elements are needed in order to supply the population during all seasons of the year with food of adequate quality. Production by hobby gardeners never could satisfy the high demand of the population.

The food business in Germany is comparable to other industrial nations marked by division of labour, professional procedures and high quality requirements, in strict accordance to EU hygienic standards [EG 178/2002, EG 852/2004, EG 853/2004, and EG 854/2004].

Nowadays, farmers in Germany and worldwide are part of a professional supply chain, which is highly standardized and able to deliver food just-in-time under economic pressure, similar to other industrial process chains. Good logistics and high efficient transports in the food business are normally performed by specialists in this business area.

It is assumed that these characteristics especially lead to an acceptable price, both in terms of economy and of ecology. Hence, local or regional food has to compete with European-continental and with global food, as mentioned before.

A rather special role in the food business is played by so-called organic food, which is especially labelled as such food. In the history of labelling organic food, some private associations in Germany started with labels such as Bioland[2], Demeter[3], or Naturland[4]. Meanwhile governmental authorities followed up with official labels. In Germany the governmental Bio-label started in 2001, characterizing organic food, which has not been irradiated, no genetic engineering, no plant protection measures, and no lightly soluble fertilizers, in accordance to the EU decree for organic food [EWG 2092/91].

Surprisingly, the matter of energy efficiency in the entire process chain – farming, transport, and distribution – does not play any role in the labelling of organic food. Nevertheless German consumers believe that organic food automatically is associated with local or regional origin and by this means saves energy in comparison to European-continental or global food.

At the same time, scientific publications indicate the conception of regional produced food items being more ecological than global distributed consumer goods. Kranendonk e.g. is researching water and energy turnovers using fruit juices from a Brazilian origin as well. The publication reports about 22,458 kg water and 139 kg gas per ton of fruit juice. Unfortunately,

---

[1] CMA: Centrale Marketing-Gesellschaft der Deutschen Agrarwirtschaft (Central Marketing Association of the German Agrobusiness) look at http://www.cma.de.

[2] Bioland: please look at http://www.bioland.de.

[3] Demeter: please look at http://www.Demeter.de.

[4] Naturland: please look at http://www.naturland.de.

the origin of the data and the scientific procedure of allocation remain uncertain [Kranendonk 1993].

A Swiss research project is estimating the turnover of primary energy and the release of carbon dioxide for the production and distribution of lamb meat from New Zealand. These results are compared with Swiss lamb meat. This direct comparison is made possible, because the data for the breed and slaughter of the lamb, as well as the circumstances for packaging and storage of the meat are assumed to be the same for lamb meat from Swiss and New Zealand. Only transportation data are different. Because of simple equation of Swiss and New Zealand lamb breed the conclusions are based on obviously false assumptions, and the data base remains uncertain [Zamboni 1994].

Weizsäcker uses the Kranendonk example - orange juice - and indicates a consumption of 40 million liters of gas only for the transport from Brazil to Germany. He recommends the consumption of red currant juice made from regional German crop instead of orange juice made from Brazilian concentrate. This replacement would cause less energy turnover and, at the same time, provide the population with more vitamins. Besides the fact of missing German capacity of redcurrant crop throughout the year, the publication does not indicate the used scientific method and gives no empirical data for the published findings [Weizsäcker 1995].

Jungbluth provides a comprehensive overview regarding this kind of research studies. Unfortunately, there is no more specific information on case studies like fruit juices or lamb meat in this publication [Jungbluth 1998].

Due to the general assumption that regional food distribution is involved with lower energy turnovers and smaller resource inputs than global interactions, Koerber prefers regional consumer goods. In addition, he asserts lower emissions of environmental impacts and, as well, claims better economics on average, because of smaller marketing distances. Nevertheless, the paper demonstrates still a lack of specific empirical data to prove these conclusions [Koerber 1999].

Jungbluth is discussing the Swiss food consume and way of import, regarding its impact on environment [Jungbluth 2000]. In his dissertation thesis he performs a modular life cycle assessment method. The investigation is based on log books and consumer diaries. In addition, the Swiss food market and the origin of the different food items are analyzed. Calculating the impacts of food import to the Swiss market, the author comes to the conclusion, that local or regional Swiss food saves energy in all cases.

This conclusion is based on the far-reaching assumption, that the import of fresh food deriving from overseas to Switzerland is performed by plane only, bridging a distance of 4,700 km on average. But, following his own information of the official Swiss import statistics, all the import pathways demonstrate the rather negligible part of air transports, because a very few percent of food are definitely imported by plane only [Jungbluth 2000:194 f]. On the contrary, even more than 95 % of meat is transported to Switzerland by road using HGV. Hence, the results of that calculation cannot be taken seriously, but depend on strong misleading assumptions, unfortunately. Again, the central question, why food should be different from other natural goods in terms of energy impact, remains unanswered. And: the claim of "regionalism" as environmental friendly is not based upon valid assumptions.

Most interesting in the public debate is the fact, that this claim of "regionalism" is related with food only. But there are quite a lot of different goods we take from environment for human use. Just think about clothes, which are made from natural raw material such as cotton, silk, wool and linen. Surprisingly, nobody ever is talking about regional clothes!

Cellulose as raw, intermediate, and final product is deriving from wood. This material is used for clothes, nappies, handkerchiefs, and further hygienic articles for human use. Never heard about regional nappies!

Tobacco leaves, which are the raw material for producing cigarettes, cigars and pipe tobacco, are a product from nature, even perhaps of organic quality. Tobacco leaves need farming, crop, transport to facilities, production, packing and - last but not least distribution and logistics up to the point of sale. All these single process steps require energy turnover comparable to food. Tobacco plants are growing even in Germany, but never ever is someone asking for regional cigarettes because of supposed low energy turnover.

Nowadays the area of bio fuels is absolutely booming. Agriculture increasingly is involved into producing gas in order to replace fossil fuels. Rape oil, corn starch, and sugar cane are just a few examples for this increasing agribusiness. Meanwhile there even a competition can be observed between these raw materials for food and for bio fuels use. Again: Regional demand of such natural products is no object of discussion.

Furthermore, renewable raw material is used for detergents and medicines, which are as well never asked to origin from local or regional sources.

All these facts lead to the assumption that food is a very special and rather emotional good, and not comparable to clothes, bio fuels, tobacco, cellulose, detergents, or medicine. Hence, food is regarding the human emotion of the consumer not comparable with other industrial goods, where pure rational arguments are counting. This is one of the challenges for agribusiness and food trade, because there is a gap between the emotional attitude and expectations of the final consumers on the one hand and the rational reality in the business area on the other hand. From the scientific point of view, which should be rather independent from emotional access, the question of energy economy in the food related process chains needs elaborated attention by empirical analysis.

Energy Carriers in the Food Business

As a matter of fact, all food process chains if local food, regional food, European-continental food or perhaps global food require energy turnover. At first, energy is needed for farming, crop, and breed. This includes e.g. power-feed delivery and plant protection measures. Even the efforts of shepherds, cowboys or the veterinary services are not possible without energy turnover, which has to be taken into account.

Energy is required for transporting the breed to the slaughterhouse or the crop to the plant processing units. The meat and plant processing facilities need energy in order to produce their intermediate or final products. All the production requires packaging, which is performed by hand or machinery. The packed products have to be transported and distributed again, until they get to the point of sale. And finally the consumers need energy in order to do their shopping and bring the final products home.

Energy for all these different purposes can be supplied as gas, diesel, natural gas, firewood, heating oil, long-distance heating, or electricity. These types of energy carriers we call delivered energy (synonym: secondary energy), which is sold to the final consumers and measured in different sales units. Table 2 presents an overview to these terms, including information about the units usual in trade and the natural sources, which we call primary energy.

It is important to know, that most of these energy carriers are worldwide comparable in terms of greenhouse impact. It does not really matter, at which place of the world heating oil, firewood, crude oil, kerosene, diesel, gas and natural gas are burnt in order to provide the user with heating or drives. All delivered energy carriers emit gaseous carbon dioxide (CO_2) and water (H_2O) as final products of combustion. The rates of emission do not depend on the place of combustion, but depend on the type of energy carrier only. Looking at converted sources of energy as long-distance heating and electrical power things are quite different, because the rates of emission in these cases depend on the underlying energy mix of each country.

In order to compare different energy carriers in terms of greenhouse impact, at first the sales units have to be converted into the physical energy unit kilowatt hours [kWh]. Based on these unit conversions, the specific emission of carbon dioxide in g/kWh can be calculated.

Table 2. Important energy sources in agribusiness and food process chains

Delivered energy	Sales unit	Common use	Primary energy source
Heating oil	liters [l]	heating	mineral oil
Firewood	solid cubic meter [sm^3]	heating	forest wood
Heavy fuel oil	tons [t]	vessel drives, power stations	mineral oil
Kerosene	tons [t]	jet turbines	mineral oil
Aviation Gasolin (AVGAS), Die-sel	liters [l]	airplane, HGV, trucks	mineral oil
Gas	liters [l]	cars	mineral oil
Natural gas	cubic meter [m^3]	heating, drives	natural gas
Long-distance heating	kilowatt hours [kWh]	heating	fossil or waste fired communal heating and power stations (CHP)
Electric power	kilowatt hours [kWh]	drives, lighting, process heating	energy mix of nuclear power, fossil fired power plants, hydro- and wind power, solar panels

To convert the sales unit into energy units, the specific heating values of the different energy carriers are to take into calculation. Table 3 presents the conversion into energy units and the different specific release of carbon dioxide (CO_2) of the named energy carriers.

All data of table 3 are rounded up or off as to give an overview regarding the different energy values of the differing energy carriers. Real energy carriers may deviate a little bit because of the natural and industrial variety of energy products. Natural gas for example varies in its composition, depending on the different natural deposits. The composition esp. regarding the methane concentration influences the heating value of natural gas. But for the purpose of calculating the ecological impact the precision of these conversions should be sufficient, because life cycle assessment (LCA) in all cases is a comparing method, which should identify significant differences in pathways or process chains. For that reason the precision of the information of table 3 is adequate.

Table 3. Energy carriers in terms of energy units and specific CO2-release

Delivered energy	Sales unit	Conversion into kWh	Specific CO_2-release [g per kWh]
Heating oil	1 Liter [l]	1 l = 10,4 kWh	260
Firewood	1 solid cubic meter [sm^3] = 400 – 500 kg	1 kg = 4 – 4.5 kWh, 1 sm^3 = 1,600 – 2,250 kWh[5]	0[6]
Heavy fuel oil	1 ton [t] = 1,000 kg	1 t = 11,050 kWh	260
Kerosene	1 ton [t] = 1,000 kg	1 t = 11,900 kWh	260
AVGAS, Diesel	1 Liter [l]	1 l = 9.8 kWh	260
Gas	1 Liter [l]	1 l = 8.9 kWh	260
Natural gas	1 cubic meter [m^3] = 1,000 l	1 m^3 = 10 kWh	200
Long-distance heating	1 kilowatt hour [kWh]	1 kWh = 1 kWh	300 – 400 (depending upon the on-site situation)
Electric power	1 kilowatt hour [kWh]	1 kWh = 1 kWh	560 (German average)

As mentioned already, the conversion of long-distance heating and of electric power units into primary energy and specific CO_2-release depends on the national or even regional energy mix. Some countries like e.g. Austria, Brazil, New Zealand and Norway are able to produce electricity by hydropower. Therefore the CO_2-release of these countries is remarkable low. Other countries like France use nuclear power stations to a high extent, causing low CO_2-emissions as well. The data of table 3 regarding long-distance heating and electric power are related to Germany, only. Worldwide investigation of process chains has to keep in mind, that especially with electricity and long-distance heating differing emission factors in different countries are to consider.

Central Hypothesis: The Ecology of Scale

In consideration of the given examples in the non-food area and the obvious preconceptions regarding "regionalism" in the food business we scientifically follow the assumption that not only transportation and distribution distances but also efficiency and sophisticated logistics of the whole process chain are decisive for the specific energy turnover of different food items. A global acting enterprise could possibly be more efficient in terms of energy than a local farmer, in spite of global distances.

---

[5] Conversion into energy unit actually depends on wood species and its humidity.

It is assumed, that the specific energy turnover, which is related on the functional unit, depends negatively on the business size. We expect a more or less decreasing function with increasing business size.

This hypothesis has been formed by comparing food with non-food items, due to the fact that, e.g., clothes, tobacco, wood, detergents and further goods from natural sources are always produced and distributed regarding only the economic and in some terms the ecological point of view. Regional aspects do not play any role in all business areas except the food business.

From economy we know, that the specific financial effort is depending upon the number of items which are produced, because of the fixed costs to be taken into account. This well-known law of economy is called *"The Economy of Scale"*. This term is known worldwide and accepted by the scientific community, although the financial efforts per item only describe just one small aspect of the entire business economy!

Regarding the ecology, we actually assume the same regularity as in economics: Above all, the specific ecological impact should be dependent on the number of produced items. We assume small business units to be less effective in terms of ecology compared with bigger units. This central hypothesis we call *"The Ecology of Scale"*, keeping in mind, that energy is only one aspect of ecology, even though one of the most important aspects.

The energy turnover of process chains, which are assembled by many participants – starting at the primary production in agriculture and ending up at the point of sale, determines the CO_2-release, the Global Warming Potential (GWP) and the Ozone Depletion Potential (ODP), among others.

Additionally, the emission of nitrous gases or sulphur oxides, fine particulate and other traffic emissions are related with energy turnover as well. Even noise originating by vessels, trains, trucks, and planes is proportionally connected with energy turnover. Regarding the transport of power feed, plant protection measures, or medicine to the farms, the ecological impact again is close related with the entire energy turnover. In all these respects the term of "Ecology of Scale", which has been created in the context of our present research, pools many ecological impacts, measured by the energy turnover as leading unit.

In further development of our central hypothesis called *"Ecology of Scale"* we expect the opportunity of benchmarking different process chains in terms of energy. As mentioned already, the specific energy turnover will be shown in relation to the business size. By comparing these coordinates, it should be possible to define a minimum business size similar to a *"break even point"* in economy. This *"break even point"* should determine the least size of a process chain, which is needed to be competitive in terms of energy turnover. But, regarding the complicated matters of process chains in the food business, we would rather expect an area and not just one point, which will lead to a minimum business size, required for competitive units in terms of energy. This topic is here characterized as second hypothesis, which is to prove by scientific investigation.

From this point of view and based on the central hypothesis of "Ecology of Scale" the objectives of our scientific field studies are as follows:

- Identification of food examples as to compare process chains of different size,

[6] Zero release is here reasonable, because firewood is considered to be a renewable energy carrier, which rebinds CO_2 from atmosphere while growing.

- Investigation of the turnover of delivered energy at each single process step,
- Allocation of the entire energy turnover to the business size,
- Falsification or verification of our central hypothesis and
- Assessment of reasonable break even areas and benchmarking for case studies and food examples.

At the same time, it is quite important to investigate, if the origin of the food – whether local, regional, European-continental or global food – determines the specific energy turnover or not. The assumption of an ecological advantage regarding food of local or regional origin compared to food of European-continental or global origin has scientifically to be verified or falsified as well. The question is if the transport distance as such or rather the means of transportation, including the aspect of distance will be decisive for the specific energy turnover of the entire process chains. As showed above, the trade and consumption of food from local or regional crop and breed is recommended by many politicians and by some scientists in Germany, although the conclusions are not data based on scientific investigations.

Method of Research

To verify or falsify the named hypothesis of "Ecology of Scale" by empirical data and further to improve the used method of Life Cycle Assessment, a scientific evaluation of specific energy turnover of food production and distribution is applied, comparing local and regional with European-continental and global food processes. The purpose of our data research is to observe whether the operation efficiency and the logistics are more important than the transport distances as such, regarding all the specific energy efforts of the whole process chain.

Using the energy component of Life Cycle Assessment (LCA) as standardized method the energy efforts of the whole process are to identify and then to allocate to the food items as functional units. LCA is defined as a comparison of the environmental impacts of two or more different but comparable products, processes or systems [UBA 1992]. The method is meanwhile published as international standard [DIN EN ISO 14040]. Parts of LCA are energy balances. The turnover of energy in all steps of the process is to evaluate first and then to allocate to the functional unit. Primary energy and environmental impacts like carbon dioxide emission can be calculated from this data base.

At first, food examples have to be identified as to perform case studies. All food examples chosen are to consider as local or regional and European-continental or even global process chains. And, the food examples should cover a broad range of different food items, regarding vegetables, meats and even drinks, if possible. From the LCA-point of view, single and homogeneous food products are more suitable for such an investigation compared with convenience food composed of different heterogeneous items. As a matter of fact, to investigate a complicated process chain of food we recommend to look after e.g. apples or lamb chops and not to observe more complicate products like vegetable soups or fast-food meals ready-to-eat. By the way, the described claim for "regionalism" is rather applied for such primary products like apples, wine or lamb chops and not for more convenient meals ready-to-eat. In that sense, our selection of food examples for investigation is conforming to the idea of "regionalism", because nobody asks for local or regional origin of fast food!

As second step, after deciding the food example, all the single steps of an entire process chain have to be identified. Usually, the investigation starts at the point of sale. Here, the retailers of the defined food items are to ask for their specific suppliers. The second step identifies the wholesale trade of that food. By this means, more and more steps of the process chain can be identified. Finally, we come to the primary production and agriculture, wherever that will be. All intermediate steps of production and trade have to be included, in order to complete the entire process chain and to identify all actors. A schematic diagram of an entire process chain is presented in Figure 1.

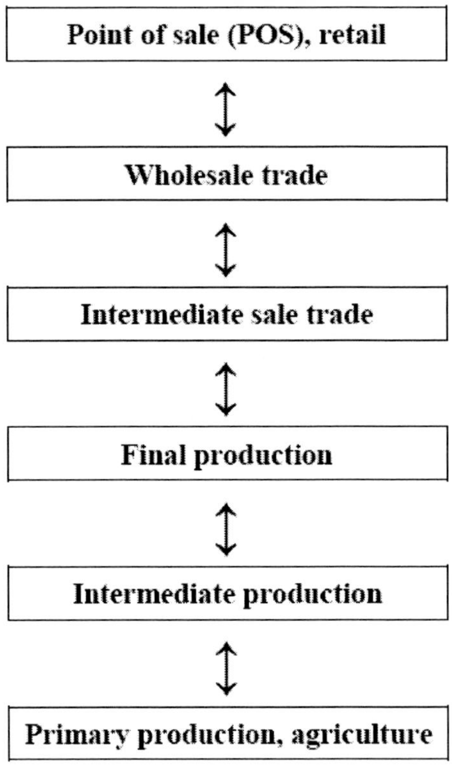

Figure 1. Schematic diagram of an entire food process chain.

The participants of primary, intermediate and final production as well as the different actors of trade are connected by distribution and transport. Transport in our sense means uni- or bi-directional transfer of goods from one actor to another part of the process chain. In all cases the trip back to the place of departure has to be taken into account, if there is no backload.

Distribution in our sense is defined here as sum of transports from one part of the process chain into the field. For example, if wine in barrels is transferred from the vineyard to a bottling unit at some other place, we call this effort as transport. If the bottled wine then is transferred to many retailers within the whole country, we talk about distribution. Usually, distribution can be analyzed by macro-analysis, asking for the entire amount of distributed food and the efforts in terms of energy, related to this amount.

A second example for the definitions of the terms transport and distribution in our sense is lamb meat. That food is transported from New Zealand to the port of Hamburg, Germany and distributed from the port of Hamburg via some intermediate sale trading posts to many retailers all over in Germany.

Transport and distribution in some cases does not belong to the core business of farms or of the food production units. In these cases, all the additional transport companies have to be identified, in order to later find out and allocate their energy turnover as well. In other cases, the needed transports of raw, intermediate and final products are performed by the production unit itself. Even mixed procedures are possible, if some part of the production is transported by the production unit itself and some other part by haulage contractors. As a matter of fact, it is quite important to know all these details about the means of transportation, the distances and the energy consumption of the transport units. A schematic diagram of an entire process chain with external transport and distribution companies is given in Figure 2.

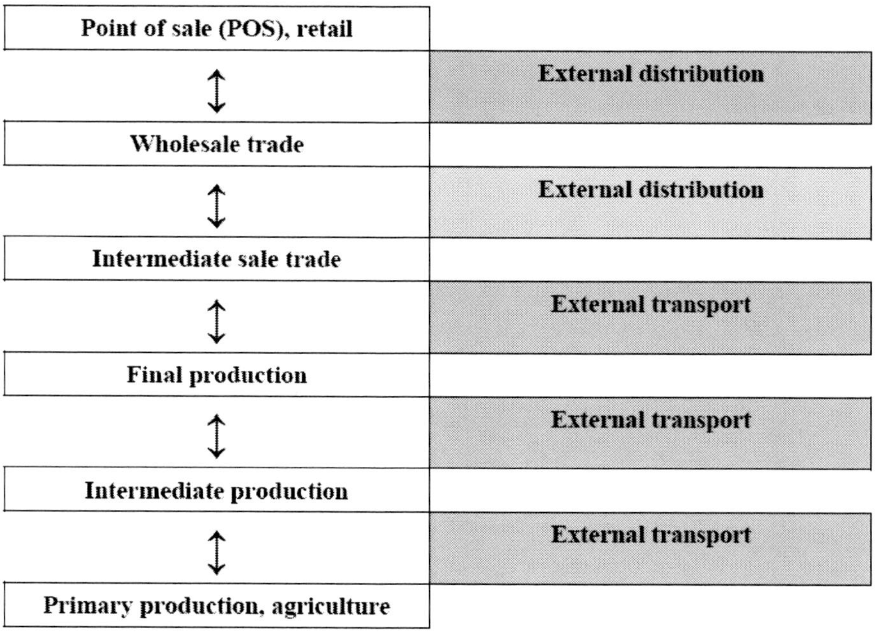

Figure 2. Schematic diagram of food process chain including external transport and distribution companies.

Figure 2 represents quite a number of different business units organizing an operating process chain. In that sense, it does not matter, if the process chain is operating in a local, regional, European-continental or even global area. In all cases, there are many actors all over the world as participants of food delivery to the final consumers, who buy food at the point of sale in Germany. Each business unit, which belongs to an operating food process chain, is a subsystem of the entire process, and has to be identified and investigated.

Obviously, Life Cycle Assessment (LCA) requires system boundaries, as to keep the results comparable. In our studies the investigation begins at the primary production in the agricultural area. Pre-operating units such as feed production in the area of breeding are included just in case the feed production for the livestock belongs to the breeding farm itself.

In case, the feed is bought by the farmer from another company or farm, which is a specialist in crop farming, the efforts to produce the feed are out of the present investigation area. But, if the transport of the feed to the breeding farm is performed by the breeding farm itself, this energy turnover will be included so far.

A second example for system boundaries is the energy turnover for plant protection measures. The hidden energy of plant protection agents, which are produced by some agrochemical companies operating external, is not included into the present investigation. On the other hand, the energy turnover used for transport and application of these agents onto the field is included into the investigation system, because this is usually performed by the farmer himself.

The same modality we have with energy turnover for producing pharmaceutical measures for the livestock of the farm, with energy turnover for manufacturing machines, tractors, cars, trucks, HGV, vessels, aero planes and for all the secondary equipment necessary for the entire process chain. All external and hidden turnover of energy is definitely not included into our system investigation. This procedure and modality is in accordance with LCA and by the way quite understandable in order to limit the extent of the investigations.

Additionally, the question would be open, where to stop with the implementation of all external energetically factors? In case, all external and hidden energy should be considered, then, for example, the energy for manufacturing of the "golden buttons" of the captain who is controlling the global operating vessel had to be taken into account as well. To make things clear, external energy turnover basically is not included into the system and the subsystems investigated here.

The present research ends up at the point of sale. Hence, the energetic efforts of final consumers are not included into our research. The question of how to most efficiently manage the needed transports from the point of sale to the final consumers in the private households is still open. First of all, the consumers go shopping, and doing that they need transport measures to bring themselves to the point of sale. Secondly, the final consumers need energy carriers of different kind in order to bring themselves including the purchase back home.

In Germany, most consumers take their own private car to get to the shopping mart. But, some suppliers deliver their food to the consumers directly, while the purchase is ordered by phone or internet. Only a few consumers are able to walk to the food market because of a rather small distance between home and market.

These few remarks demonstrate the individuality of consumers behavior, which would have to be considered in terms of energy turnover for shopping and food delivery to the point of consume at home. It would be quite interesting to investigate this question. Probably, the answer is to find by macro-statistical analysis only, and not by individual questionnaire.

As mentioned already, the present data research is focusing on the food business and its process chains, in order to compare different business size and different places of food origin. This investigation ends at the point of sale, whereas the individual procedures of going to the market and getting back home with the purchased food are out of observation. Furthermore it can be assumed, that the excluded part of consumer behavior does not influence the investigated question of the food origin.

Nowadays, it is to observe, that food of all kind whether local or not, whether organic or not is marketed by the same manner. Most of the food is put to trade by supermarkets and discount retailers. Only a very small part of food is traded by weekly markets downtown. A survey on a weekly market downtown will demonstrate that the food sold there is coming from all over the world: apples from New Zealand, broccoli from Ecuador, flowers from Israel or Kenya are just beside fruits of local origin, depending on the season of course. Hence, the procedure of not including the consumer behavior into the investigation of energy turnover of food process chains is correct in accordance with Life Cycle Assessment, because this final part of the process chain seems to be identical regarding food of different origin.

An important objective of Life Cycle Assessment procedures is the comparison of different pathways which are accessible for the same purpose. Due to the aspect of comparison all assumptions must be declared. And, all assumptions must be valid for all different pathways. Hence, the described assumptions above are definitely valid for food of local, regional, European-continental or global origin. The energy turnover exterior of the food process chains is not taken into account as declared above.

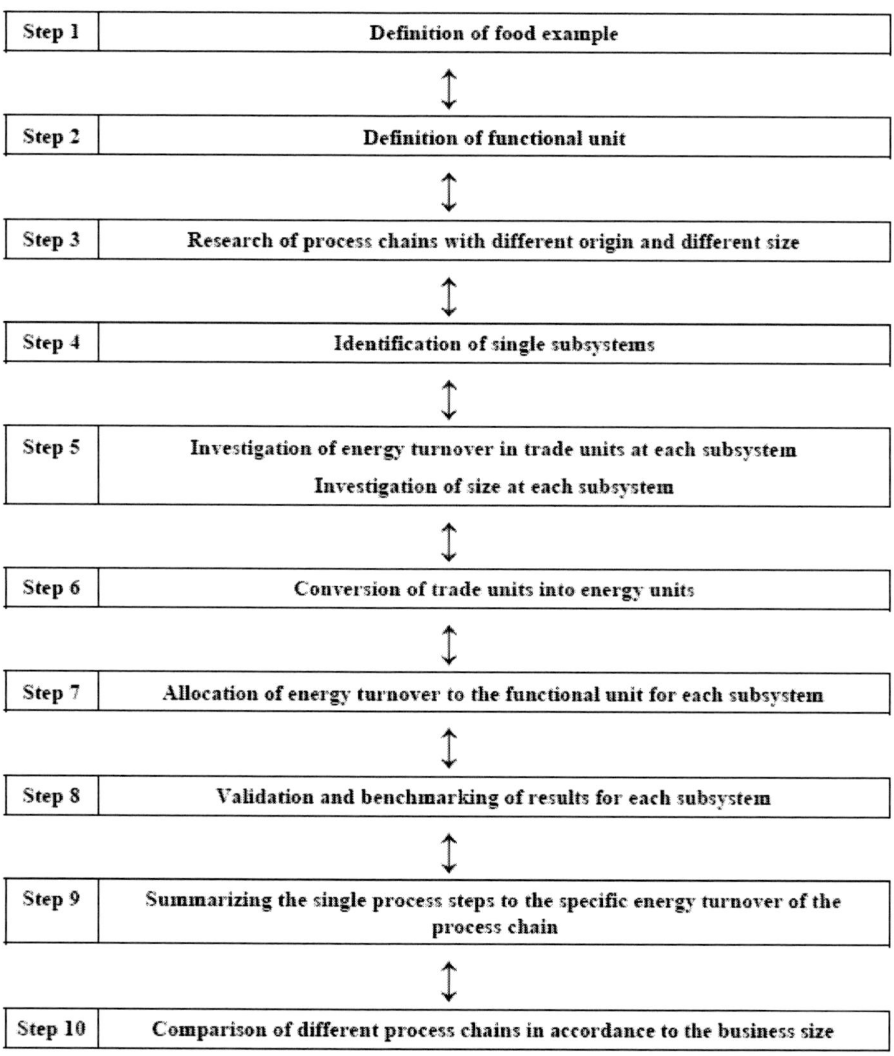

Figure 3. Methodical procedure of the scientific research.

In order to compare food process chains of different size, it is necessary to define functional units as reference quantity in a process chain. All subsystems of the researched process chain are evaluated by means of the same functional unit. Obviously, the functional unit within one food example must be the same, not depending upon the question of origin. Food of local, regional, European-continental or global origin has to be evaluated by means of the same functional unit. In that sense, the definition of suitable functional units is part of the methodical progress in this scientific research.

The methodical procedure of the present research is basically presented in Figure 3.

Obviously, the scientific method of our investigation is qualitative field research, performed by empirical investigation. The main research topic, as to compare different process chains and business units of differing size, demands the application of this qualitative method.

As a matter of fact, quantitative scientific methods would not answer the actual questions of this research project. As a part of LCA, the energy turnover of comparable units is investigated, by personal interview worldwide (see step 5 of Figure 3). All the interviews are prepared by a qualitative questionnaire. The presented answers of the business managers are supported by copies of the energy invoices and former business data sheets.

For the purpose of validation it is important to evaluate the answers by further questions. For example, data of the yearly gas consumption of a truck can be supported by additional data of the type of truck and the yearly mileage. This example is figured out in table 4 and table 5.

Table 4. Questions, answers and 1st data assessment (example)

Step	Question	Answer, result
1st question:	Consumption of gas of a specific truck per year	2,400 liters
2nd question:	food load of the truck	3,500 kg
3rd question:	frequency of transports	weekly = 52 times per year
Calculation step 1:	Consumption of energy	2,400 l * 8,9 kWh/l[7] = 21,360 kWh
Calculation step 2:	Food transport per year	52 * 3,500 kg = 182,000 kg
Result of allocation:	Specific energy turnover	21,360 kWh / 182,000 kg = 0.12 kWh/kg

Table 5. Evaluation of data assessment by further questions (example)

Step	Question	Answer, result
4th question:	Type of truck	Light truck for food transport
5th question:	mileage per trip (one way)	100 km
6th question:	miles per gallon	10 mpg = 4.2 km per liter[8]
Calculation step 3:	mileage per year	52 weeks * 100 km * 2[9] = 10,400 km
Evaluation:	Estimated gas consumption	2,476.19 liter

[7] Conversion by heating value of gas; see table 3.
[8] Unit conversion: 1 gallon = 3.78 liter; 1 mile = 1.6 km; 1 mpg = 0.42 km per liter.
[9] Multiplication with factor 2 because of trip back without load.

The final allocation in the given example is based upon the data received by personal interview, using the answers to the first questions. Further questions are necessary as to evaluate the former answers. In the present case, additional questions for the type of truck, mileage per trip and miles per gallon are quite useful.

The given example demonstrates the pretty good coincidence of the answer presented to the 1st question with the evaluation shown in the last line of the table. Hence, the result in terms of energy turnover per functional unit calculated in the given example to 0.12 kWh/kg is validated for this single step of the process chain.

A validation like this has to be performed at each step of the entire process chain. Obviously, this qualitative research is taking many resources and needs a lot of time, especially for global and continental food. In case, primary data cannot be validated, further explanations and investigations are necessary. In all cases, it is recommendable for any researcher, to visit each process unit or subsystem himself with open eyes, in order to get an appropriate impression of that specific part of a process chain, whether of global or of continental or of regional or of local origin. In science, this is the only way not to be prejudiced by personal opinion or by public view.

Case Studies

In order to evaluate the scientific hypothesis of Ecology of Scale, qualitative case studies are performed. The case studies investigate significant examples of the food business. At first, food examples have to be designed, as to compare business units of different origin and different size.

As mentioned already the examples for the case studies should be consisting of homogeneous food items. Heterogeneous food items which are composed by different parts in industrial food processing units are not suitable for this type of scientific research.

The case studies performed within the last ten years of research are looking at food examples as follows:

- Apples,
- Beef,
- Fruit juice,
- Lamb meat,
- Pork and
- Wine.

With that selection of case studies a suitable range of different food items is covered.

Table 6 gives an overview to the selection of case studies.

These food examples represent food of vegetable origin as well as meat products. The application of food processing after the crop or breed is quite different as shown in table7.

As demonstrated in table 7, one of the food examples fruit juice normally requires intensive processing, which is usually performed in processing facilities of industrial size.

Table 6. Selection of case studies: origin and sales unit

Food example	Places of origin	Sales unit and package
Apples	South Africa, Germany	1 kg bag
Beef	Argentina, Germany	1 kg vacuum bag
Fruit juice	Brazil, Europe, Germany	1 liter glass bottle (returnable)
Lamb meat	New Zealand, Ireland, Germany	1 kg vacuum bag
Pork	Hungary, Germany	1 kg vacuum bag
Wine	South Africa, Hungary, Germany	0.75 liter glass bottle (non returnable)

Table 7. Selection of case studies: types of food and food processing

Food example	Type of food	Type of food processing
Apples	Vegetable crop	Extensive, because no processing but storage required
Beef	Meat products	Intensive by handwork (slaughter) or semi-industrial manufacturing
Fruit juice	Vegetable origin	Intensive food processing by industrial manufacturing
Lamb meat	Meat products	Intensive by handwork (slaughter) or semi-industrial manufacturing
Pork	Meat products	Intensive by handwork (slaughter) or semi-industrial manufacturing
Wine	Vegetable origin	Intensive by handwork (wine producers) or semi-industrial manufacturing

Other case studies focus more or less on handwork or semi-industrial manufacturing, which is performed by rather small and medium sized facilities. Apples are the example of extensive vegetable crop, usually performed by handwork, if fruits of high quality for personal consume are to earn. Apples do not underlie any further handwork or industrial processing. But, on the other hand, this fruit may require highly sophisticated storage units, some of them with controlled atmosphere in order to cover typical saisonal effects.

In sum, the selection of suitable case studies looks after the suitability on the one hand. On the other hand, different types of food, package and processing intensities are useful, in order to cover a broad range of food items. Finally, the places of origin and the business size of the process chains should be different as to compare process chains of different origin and size.

The following chapters will give more detailed information about the food items in the case studies and present the primary data as results.

Apples

The consumption of apples in Germany is rather high. Altogether we find a trade of 1.5 Million tons apples per year, according to a yearly demand per capita of around 18 kg. In that number the further demand of the juice industry is not included. Apples are by far the most popular fruits in Germany [StaBu 2007].

The annual harvest of apple in Germany amounts about 850,000 tons [STABU 2007], a part of which is used for producing apple juice. The self-sufficiency of Germany with apples for direct consumption can be assumed to round about 50 %. Hence, Germany is highly reliant on importing apple from other countries, mostly from Italy, France and Poland. Additionally, apples are considerably imported from South Africa and New Zealand, dependent on the season of the year.

As mentioned already, about 50 % of the demand can be satisfied from German regions. The most important regions of apple production in Germany are the region around Lake Constance, some parts of Rhineland near Mainz and a region close to Hamburg called "Altes Land", beside others. The earning and crop takes place in August, September and October of the year, due to climate conditions.

The entire crop of German regions is not sufficient to cover the yearly demand of the population, although high sophisticated storage facilities under controlled atmosphere (CA) are stretching the season. Using this storage technology, apples originating from German regions can be supplied from the time of crop (August to September) until April next year. This is corresponding to more-or-less half a year of storage under controlled conditions.

CA-facilities store apples under cooling conditions and reduced oxygen atmosphere. The time of ripening is controlled by increasing the temperature and the oxygen content of the atmosphere. The maximum time of storage depends on the type of apple and the actual quality of the yearly crop, of course. The storage technology of controlled atmosphere (CA) is relevant for the question of specific energy turnover, because cooling and modifying the atmosphere require energy. From this point of view, the process step "CA-storage" has to be considered.

On the other hand, German retailers are reliant upon import of apples from European-continental and global origin, because the German apple production is not sufficient, both in terms of turnout as well as in terms of seasons. The rate of import amounts some 50 % on average. The most considerable countries of apple import to Germany are The Netherlands and Italy [StaBU 2007].

But, regarding these governmental statistics, it is to remark, that Italy is indeed an important European producer of apples. The Netherlands as rather small country themselves import most of the fruits to the European Union (EU). So, the Netherlands serve as EU-frontier port, especially via Rotterdam. This important role of The Netherlands falsifies German statistics of fruit import: Import from The Netherlands does not mean Dutch apple production of the same amount. Hence, the governmental statistics are misleading in this case.

At the port of Rotterdam (Europoort) operates a high specialized fruit terminal providing the import/export trade with loading and unloading facilities as to easily combine vessel and HGV carriers. At the same time, Europoort-Rotterdam operates large cooling storage and ripening facilities with CA-technology, located very close to the quayside. By means of this quite efficient logistics, Europoort-Rotterdam delivers any kind and amount of fruits to Germany, just-in-time as requested by German wholesale trade and retailers.

Global producers with enhancing interest for Germany are New Zealand, South Africa and Argentina. Meanwhile 4 % of the apple consumption in Germany is deriving from New Zealand, further 4 % coming from South Africa and Argentina. All these global imports are shipped by freight vessels, mainly to Rotterdam (NL) and Antwerp (Belgium). At present, some shipping companies still use special fruit vessels with cooling and CA-storage equipment. But, increasingly, modified container technology with standardized reefers is performed by the shipping lines, because of better logistics and higher efficiency. This progress will be enforced by economy in near future, without any doubt.

In terms of energy it is important to know, that transport of apple crop via airplane cargo to Germany does not play any role in this business, because this kind of transportation would be much to expensive. Additionally, air transportation is not necessary in regard of the fruit quality. Apples can be stored quite good and easily, and the stability is excellent in case of good storage conditions.

As mentioned already, the seaports operate sufficient cooling, storage, and ripening facilities, under turnover of delivered energy. All these aspects have to be taken into account, asking for the specific energy turnover of such a process chain.

Actually, the research includes apple farms in three main German regions: the regions of Lake Constance, of the Rhineland and of the "Alte Land". Up to now, data of these regions can be compared with global apple delivery from South Africa. Simultaneously to the energy data, the investigation covers the farms' characteristic variables as for example acreage, yearly yield and machinery.

The region of *Lake Constance* is situated in the south of Baden-Württemberg, and the largest apple cultivation area in Germany. Approximately 1,600 farms manage about 7,000 hectare of land. This acreage produces an average yield of 250,000 tons per year. This is almost 30 % of the entire German apple production. More than 90 % of the plantations follow the guidelines for integrated production. The majority of production is accredited against EurepGap[10] and against QS[11]. Since early 2004 the fruit wholesalers, which have accreditations against DIN-EN ISO 9000[12] are also accredited against IFS[13].

The apples are produced on plantations with rather small trees usually planted in single or double rows. On every farm are available several tractors. Those cover the greatest distance when spraying the trees. In the summer the trees have to be sprayed several times a month. At this, the frequency is independent from the amount of apples on the trees.

Transportation of harvested apples from the plantation to the storage facility on the farm respectively at the wholesaler is in most cases no longer than 25 km. This transport is done by the farmers themselves with in-house vehicles. In one case the apples are transported to the wholesaler by a truckage company or with a rented truck. Also, those farms which have their own storage facility need to store part of their harvest at the wholesaler. All in-house storage facilities have CA technology. At the wholesaler, the majority of apples are stored in CA storage facilities. Before storage, the apples are graded for different quality categories and packed in big or small boxes.

In order to calculate the specific energy turnover, one kg of apple crop is defined as functional unit. So, the specific energy turnover is given in kWh per kg according to reference base. The energetic effort of transport from Lake Constance to the point of sale amounts 0.15 kWh per kg. In accordance to Life Cycle Assessment, this value also refers to one year of operation. As to cover a broad range of apple crop, the data of yearly apple crop are given in a logarithmic scale. Figure 4 presents the data of the specific energy turnover of production

[10] EurepGap: Euro-Retailer Produce Working Group Good Agricultural Practice. This is an accreditation system of agricultural primary production, focusing on fruit and vegetables.

[11] QS: Quality safety is a label responding to a governmental German system of vertical accreditation, regarding agricultural products across every stage of marketing.

[12] ISO 9000 (2000): International standard for process oriented quality management systems and their accreditation.

[13] IFS: International Food Standard. Accreditation for producers of retailer branded food products.

plus transport versus the apple crop per year, as investigated at the German region of Lake Constance. In sum, seven investigated farms present data.

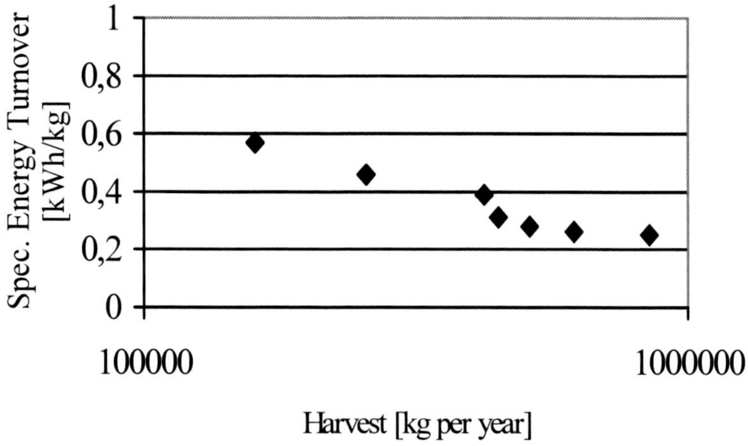

Figure 4. Specific energy turnover of apple farms at the German region Lake Constance.

The investigated data show a definite declining trend in this case. The specific energy turnover in kWh/kg significantly declines as the value of the reference base increases, as predicted by the theory of Ecology of Scale.

The apple region of *Rhineland* is located close to Mainz and Frankfurt am Main. About 1,100 farms operate some 1,800 hectare, earning per year about 21 % of the entire German apple crop. The operation conditions of the Rhineland farms are quite similar to the region of Lake Constance. The regional marketing is concentrating on a local label, which is supported by governmental authorities, in order to promote regional production and trade. The label is licensed by the ministry of economy.

Figure 5 presents the data of the specific energy turnover versus the apple crop per year, as investigated at the German region of Rhineland. The energetic effort of transport to the point of sale is measured to 0.15 kWh per kg. Five farms, investigated in this region, present their data of specific energy turnover versus harvest per year in a logarithmic scale.

The investigated data again show a definite declining trend in this case. The specific energy turnover in kWh/kg significantly declines as the value of the reference base increases, as predicted by the theory of Ecology of Scale. Just one farm, marked in Figure 5 by ■, is operating a private CA-storage facility. Hence the specific energy turnover is significant higher, compared to the other farms, as can clearly be seen in Figure 5.

The apple region of *"Alte Land"* is located close to Hamburg on the left shore of river Elbe. About 755 farms operate some 8,800 hectare, earning per year about 30 % of the entire German apple crop. The operation conditions of the farms are quite similar to the other investigated German regions. The regional marketing is concentrating on cooperative trade. Just 10 % of the yearly harvest is sold directly via farm shops to the final consumers.

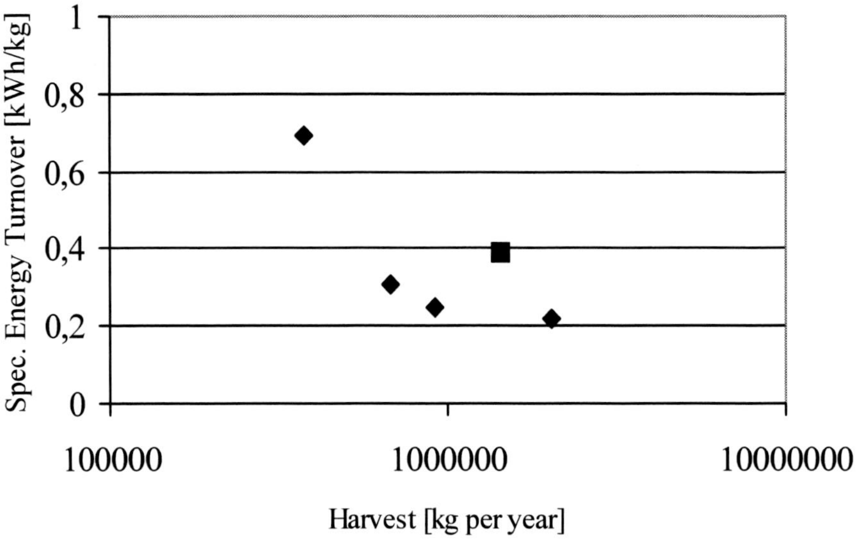

Figure 5. Specific energy turnover of apple farms at the German region Rhineland.

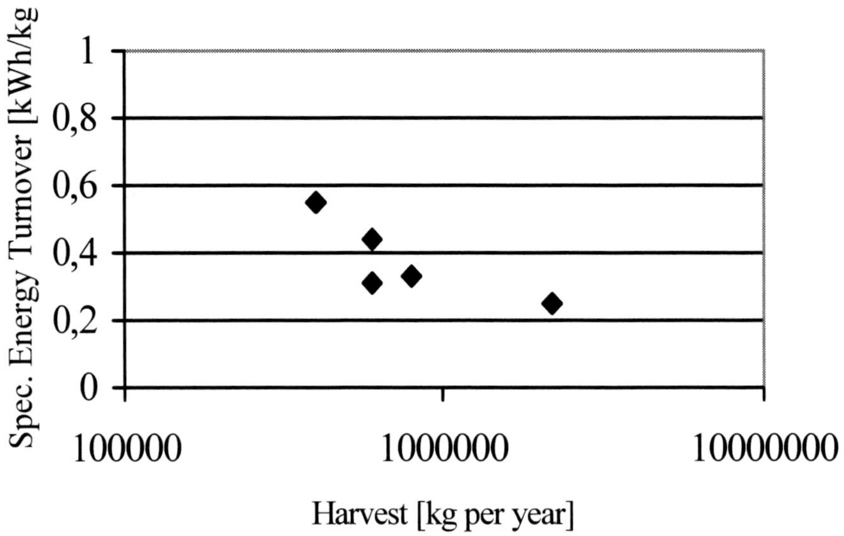

Figure 6. Specific energy turnover of apple farms at the German region Altes Land.

Figure 6 presents the data of the specific energy turnover versus the apple crop per year, as investigated at the German region of Altes Land. The efforts of transport to the point of sale amount 0.15 kWh per kg. Five farms, investigated in this region, present their evaluated data of specific energy turnover versus harvest per year in a logarithmic scale.

The investigated data again show a definite declining trend for this region. The specific energy turnover in kWh/kg significantly declines as the value of the reference base increases, as predicted by the theory of Ecology of Scale.

South Africa is enhancing its fruit production. Because this country is situated on the Southern globe, the South-African seasons fit to the lack of fruits in the Northern hemisphere during spring time and early summer. South-African farms are usually quite big and not comparable to German farms. All the fruit production is located in the region of Cape Town or close by.

A typical farm operates about 600 hectare, earning about 25,000 tons of fruits per year. The energy turnover is focusing both on gas used for trucks and tractors as well as electricity used for operating the machinery for classification and packaging. Altogether, we find a gas consumption of 405 liters and electricity use of 120 kWh per hectare of the investigated South-African farm. The specific energy turnover allocated to the entire production of fruits per year is calculated to 0.089 kWh per kg. Table 8 shows the detailed data and the allocation procedure for a typical South-African apple farm.

The data of table 8 include all internal transports of the harvest. The specific energy turnover of the South-African farm investigated here is representative for comparable farms of that size.

The external transport of the packed apple is bridging a distance of about 180 km to the South-African port of Cape Town. This transport is performed by HGV. The transport over-seas by vessel goes from South-Africa to the Europoort Rotterdam over a distance of 11,545 km.

Table 8. Primary data and allocation procedure for South-African apple farm

Process step	Question	Result
Farm	Acreage	596 hectare
Farm	Fruit tonnage	25,392.9 tons per year
Farm	Specific use of gas	404.7 liter = 3,601.83 kWh per hectare per year
Farm	Specific use of electricity	120 kWh per hectare per year
Farm	Specific use of energy	3,721.83 kWh per hectare per year
Farm	Energy in total	2,218,210.6 kWh per year
Allocation	Specific energy turnover	2,218,210.6 kWh / 25,392.9 tons = 0.087 kWh per kg

In addition, there is a distribution distance of about 500 km from Europoort Rotterdam via some wholesale stores up to the point of sale in Germany. Altogether, these three parts of the process chain are needed to market South-African apples in Germany.

The related energy effort for this part of the process chain has been investigated to about 0.38 kWh per kg. This value is valid for 2003/2004. Here, the back trip is unloaded in all cases, because specialized fruit cargo vessels and trucks are used. Meanwhile, the shipping company replaces the actual carriers by high efficient container vessels, which transport standardized reefers as fruit container on their routes. By means of standard container transports the specific energy turnover is getting much lower, because the way back is performed with backload as well. In that case, the specific energy effort for the transport from South-Africa to the point of sale is reduced to about 0.28 kWh per kg. Therefore, the entire energy turnover of production plus all internal and external transports turns out to be 0.37 kWh per kg.

This data of apple import is to compare with data of CA-storage of German harvest, because these two process chains are able to compete throughout the seasons. A good example for CA-storage can be seen in Figure 5, marked by ■. The specific energy turnover of CA-storage on the one hand and overseas transport by vessel on the other hand seems to be comparable, in terms of kWh per kg. Apples of CA-storage are burdened with about 0.4 kWh per kg. Obviously, this result is comparable to the specific energy turnover for the South-African pathway. However, more data of apple production in South-Africa, Argentina and New Zealand are required in order to further benchmarking European or German regional process chains.

Beef

Beef is one of the favorite meat products in Germany, beside pork and lamb meat. Altogether about 1 million ton of beef is consumed per year. This corresponds to about 12.4 kg beef per capita and year. Most of this food is produced in Germany, based on rather small farms which operate intensively. About 25 % of the consumed beef is imported, from European countries like France, Ireland or Denmark and from overseas, originating from countries like New Zealand or Argentina.

The local or regional process chains in Germany are usually delivering beef to a local slaughter. Hence the local pathways of beef supply to German consumers are quite easy to understand, because the local slaughter operates a butcher´s shop at the same time. Hence, the origin of the beef is even declared to the consumers. In times of BSE and other cow diseases people are quite interested to know the origin of their food.

Local process chains for beef supply are easy to investigate, because one slaughter in a small village on the countryside normally has just a few suppliers in the area around. In most cases, the farmer himself or his laborer transports the cattle from the farm to the slaughterhouse, using car or tractor with trailer. Usually, just one or two cattle are brought to the butcher. The car and trailer are going back without any load. Therefore, the gas or diesel which is needed for that transport has to be in total allocated to the beef ready to cook.

The beef producing farms in Germany operate both in grazing lands as well as in stables. In the wintertime the cattle are brought up and fattened in stables, where they get power feed. In the summertime some cattle herds are grazing outside, some other held in stables as well. Especially bulls, which are bound for meat production, are often held in stables, getting power feed all day. Only a few extensive farmers breed their cattle in the open land, taking less produce into account.

The breeding farm gets the male calves from milk producing farms, because these farms need lactating cows. The calves are separated from their mothers at once after birth, because the cows are bound for milk production. The male calves normally are a by-product of the milk farms, and sold to the breeding farms. In order to evaluate the energy turnover, all these aspects have to be investigated and allocated to the functional unit. The female calves are out of this evaluation, because the stay at the milk farms as to grow up to cows later used for milk production. In sum, the local or regional German process chains of breeding bulls are characterized by high specialized farms, cooperating with some milk producing farms at the side of entry and with just a few slaughters at the side of meat production.

This typical German process chain of local beef is presented by Figure 7:

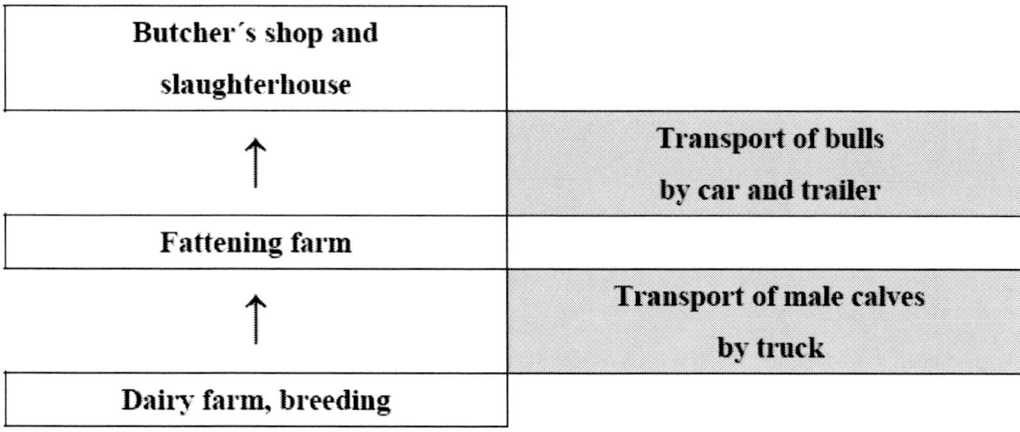

Figure 7. Local process chain of German beef production.

The international supply of beef to the German market obtains more process steps. A typical example for such a global process chain is beef from Argentina. In contrast to the local or regional beef production beef from Argentina is bred extensively. That means that the cattle herd is held outside all the year, grazing in the pampas of Argentina. The herd is supervised and managed by mounted gauchos or cowboys, who life close to the herd, using horses for their work.

Calves are born and fed by their mothers as a matter of course. Bulls, bound for beef production are driven by mounted gauchos to the central place of slaughter. As final product of the slaughterhouse the meat products are cut and packed into vacuum-bags, which are made ready to cook and suitable for gastronomy or private household consumption. Here, the cooling of the beef starts, at a temperature of + 0.5 °C.

With these conditions packed under application of vacuum and at a temperature of + 0.5 °C the beef is stable for some months and can be transported easily to Germany. The global transport of meat products is based on 20”-containers with cooling equipment, so-called reefers. This reefer is loaded at the slaughterhouse, and taken by a prime mover to the container vessel at the port of Buenos Aires. The standard container vessel follows its schedule via some further ports of destination to Europoort Rotterdam and then back home. As the vessel is fully loaded on its back trip, the energy efforts for going back to Argentina are not burdening the beef.

At Europoort Rotterdam the reefers are transshipped to prime movers and then moved to the wholesaler in Germany. From this place near Versmold (Germany) retailers are provided with beef from Argentina, which is distributed as ordered. The land-sea-land transport from the slaughterhouse in Buenos Aires to the wholesaler in Germany is highly standardized by means of 20”-container. The distribution to the point-of-sale is based on trucks with cooling equipment. All efforts of transport (prime mover in Argentina, container vessel to Rotterdam, and prime mover in Europe), and of distribution in Germany are to take into account, including the energy turnover which is needed for cooling all the time.

The characteristics of the Argentinean process chain are presented by Figure 8:

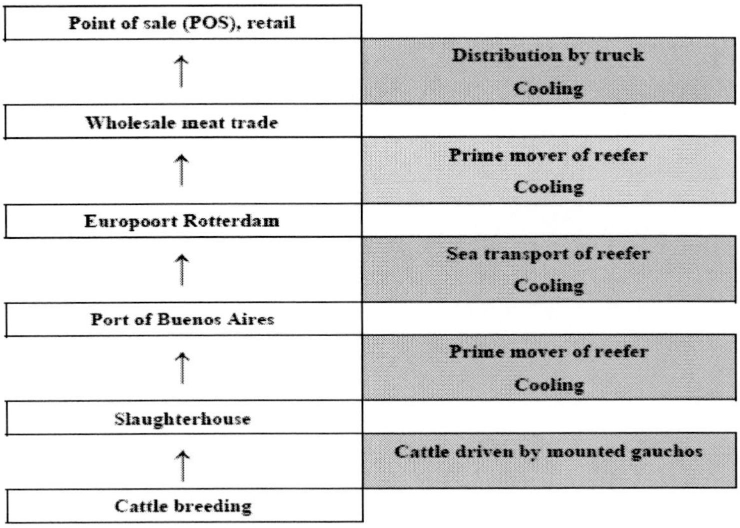

Figure 8. Global process chain of beef production and distribution (Example: Argentina).

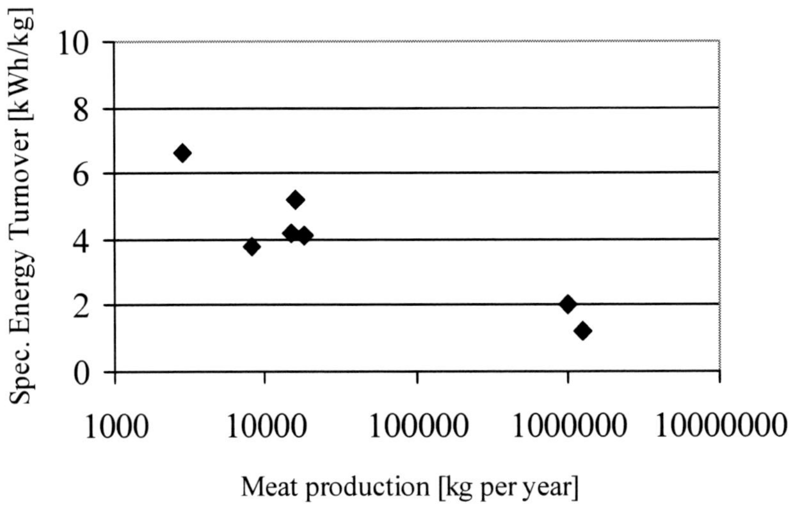

Figure 9. Specific energy turnover of local and global beef – breeding plus transport and distribution.

As a matter of fact, only pieces of high quality are supplied from global producers as Argentina or New-Zealand, for example pieces of sirloin or tenderloin, fillets or steaks. Beef which is bound for German sausages or hamburgers is usually originating from Germany and not imported from overseas.

In all cases the energy efforts of the entire process chain are allocated to the functional unit of 1 kg beef, ready-to-cook. Because slaughtering is leaving waste pieces such as bones, feet and head of the cattle, this loss of about 25 to 30 % has to be considered at performing the allocation procedure. Finally, the specific energy turnover is presented over the business size, as shown in Figure 9.

As demonstrated very clearly, small size farmers are facing severe disadvantages regarding the specific energy turnover. Although the transport distance of the local farmers could be less than 25 km per tour the energetically efforts per functional unit are significant higher than the global land-sea-land-transport of meat products from Argentina. The case study of beef gives additional evidence to the theory of "Ecology of Scale". Table 9 presents an example of calculation and allocation in order to compare local transports by car and trailer to an overseas shipping of reefers.

Local beef has no distribution efforts in terms of energy because the slaughter sells all the beef at his butcher´s shop. The slaughterhouse is located in the backyard of the butcher´s shop. This is a typical arrangement in the German countryside. But the efforts for local transport and for intensive breeding in stables burden the specific energy turnover. The global transport in reefers shipped by HGV and container freighters including cooling energy turns out to be very effective per piece despite of the very long distance.

Scale related effects are to observe in the area of slaughter. The slaughterhouse in Argentina close to Buenos Aires manages about 30,000 tons of beef per year. This amount keeps the specific energy turnover quite low, because the management and logistics is very effective. Big slaughtering and cooling units are more effective than smaller ones. The local butcher is handling some 50 bulls per year in his handwork slaughter unit.

Table 9. Indicated values of energy turnover for local and global beef

Process step	Local beef	Global beef
Fattening	2.15 – 5.3 kWh/kg	-
Transport of male calves to the breeding farm and transport of the bulls to the slaughter	0.21 – 1.77 kWh/kg	-
Slaughter	1.09 kWh/kg	0.5 - 0.1 kWh/kg
Distribution to the point-of-sale	-	1.0 – 1.5 kWh/kg
Summation	3.8 – 6.6 kWh/kg	1.2 – 2.0 kWh/kg

Table 10. Allocation of local transport to the beef (example)

Kind of data	Result
Transport distance from farm to slaughter	25 km
Tour distance including back trip	2 * 25 km = 50 km
Type of car	Pickup + trailer
Miles per gallon	10 mpg = 4.2 km per liter[14]
Gas consumption per tour	50 km / 4.2 km per liter = 11.9 liter
Energy turnover	11.9 liter * 8.9 kWh per liter[15] = 105.9 kWh
Number of cattle transported	1
Mass of cattle transported	380 kg
Rate of slaughter	58 - 60 %
Beef (ready-to-cook)	228 kg
Allocation per kg beef	105.9 kWh / 228 kg = 0.46 kWh per kg

14 Unit conversion: 1 gallon = 3.78 liter; 1 mile = 1.6 km; 1 mpg = 0.42 km per liter.

15 Heating value of gas: see tab. 3.

The cooling equipment and the butcher´s machinery need much more energy per piece compared with units of bigger size.

Scale impacts significantly the energetic efforts for transport as well. Because local German farmers transport the cattle ready to slaughter by means of car and trailer, the specific energy turnover is really high. An example for calculation based on researched data is presented in table 10:

The allocation clearly demonstrates the disadvantages of small transport units. Although the transport distance in the local area is rather small, the specific energy turnover measured in kWh per kg is quite considerable, and not negligible.

Hence, the pure distance of transport does not give any reliable information about the specific energy turnover burdened to the functional unit, because the means of transportation in relation to the distances are essential for that question. Global sea transport proves to be very effective in terms of energy, even under cooling conditions. And, air transports do not play any role in this business, because global beef is only shipped via standardized 20"-reefers by HGV and container vessels of high effectiveness. Further observation and research is under operation in order to get more data for that case study.

Fruit Juice

The investigation starts at different producers of orange and apple juices in Germany, drawing out a list of process steps for manufacturing, transportation and distribution. Fruit juice of pure quality, which is bottled into 1 liter glass bottle with returnable deposit, is either made of full concentrate, or of half concentrate, or based on direct squeezing and filling. Full or half concentrate is stabilized by concentration and can be stored for at least half a year, in order to balance seasonal effects.

Direct juice is squeezed directly from the crop and then bottled. Hence, direct juice is a seasonal product and depends on the ripeness of the fruits.

By cooperation with diluting companies, importers of orange juice concentrate (OJC) located in Rotterdam and Antwerp are included into this investigation. Step by step all the shippers via land and sea can be identified and evaluated. The final step includes the assessment of farmers and the squeezing and concentrating facilities in Brazil.

The Brazilian crop in the state of Sao Paulo is transported from the plantations to the squeezing and concentrating facilities by truck, covering a distance of 80 km on average. The primary juice is concentrated with a vacuum cross-flow rectification from 11 to finally 66 ° brix[16]. The primary squeezed juice is concentrated by this industrial-scale operation from 1 kg juice to 0.167 kg OJC. The distilled water is recovered and used for irrigation of the orange plantations. OJC of this quality can be stored and shipped easily, under application of a temperature of –10 °C and covered with gaseous nitrogen. At these storage and transport conditions, OJC is stable for about half a year.

OJC is shipped by special HGV, which are used for OJC only, to the seaport of Santos. The HGV capacity is 25 tons OJC, which is shipped over a distance of 400 km between Santos and the concentrate production plant. The sea transport from Santos to Rotterdam or

16 °brix: abbreviation of the German term „Brechungsindex“. This is a concentration unit for juices, and measured by refractometry.

Antwerp takes 13 days using special 16,000 tons vessels carrying 40 containers with 400 tons OJC each. This special type of containers is used for this purpose only.

After unloading in Rotterdam or Antwerp the loaded containers are replaced by empty ones at once. Few hours after arriving, the vessel goes immediately back to Santos under some water ballast, taking on only 11 days for return.

Calculating the vessel capacity, it is evident that one vessel carries 16,000 tons of OJC which is equivalent to 100,000 tons or 100 million liters of orange juice, after diluting. These figures illustrate that only 8 – 10 vessels per year are sufficient to provide the whole German population with orange juice. The final road transport to the German diluting companies adds up another 400 km on average, using again HGV with 25 tons OJC. The gas for all returns has to be allocated to the specific energy turnover of the juices, because all the shipping goes back without any load except some water ballast.

With this information, allocation problems of LCA can be solved easily, because we have valid data for all parts of the total system, consisting of plantations and the crop, squeezing, concentrating and re-diluting facilities including all specific transports and distributions up to the point of sale in Gießen, Germany. Cooling of the 400 tons OJC containers at the European seaports and even the return of the refilling bottles and their purification are included as well. The final results are referred to 1 liter of fruit juice [Fleissner 1999].

Secondly, data of apple juices from European-continental origin are taken into account. Here, we usually find two types of concentrate with 22 and 66 °brix, also made as well from primary juice with 11 °brix by cross-flow vacuum rectification units. Those concentrates are shipped to the diluting companies. There, purified drinking water is used to dilute the concentrates. After bottling the juice will be distributed to the points of sale. Countries of the crop origin are Poland, Great Britain, Italy and Germany.

Juices from global and European continental origin (Brazil, England, Poland, Italy, and Germany) are concentrated under supply of energy. On the other hand, the transport efforts are minimized, because of the concentration procedures before transport. From the energetic point of view, the efforts for concentrating must be balanced by the savings during the transport. Secondly, the consumers are provided with a high-grade standardized juice throughout the year, made from the origin without any additives except of high quality drinking water, which is needed for dilution in Germany.

Thirdly, we have data from local and regional farmers in Germany, who transport their apple crop to squeezing facilities inside the region. The farmers get vouchers for a part of the juice and can later pick up juices without payment from the squeezers. This signifies a cash-free system, typical for regional areas and short distances. All squeezing facilities operate in accordance to the German and European food law, following the required hygienic standards.

The researched primary data of energy turnover, which are given in tons or liters of gas, m^3 natural gas or kWh electricity, are converted into terms of energy, using their heat values. The results are allocated to the functional unit, here defined as 1 liter of fruit juice at the point of sale. Finally, the specific energy turnover in kWh/l is recorded as a function of the yearly throughput of the squeezing facility in tons/year, using a logarithmical scale covering the wide range of the investigated companies. Figure 10 shows the energy results for the production units. Figure 11 adds the energy turnover of transportations and distribution, and is therefore providing an entire overview.

In Figure 10 it is shown very clearly, that small companies with up to 100 tons of fruit squeezing per year face severe disadvantages - with a specific energy turnover range from 1.1

up to 2.5 kWh per liter. The small companies build a cluster of coordinates below 100 tons per year. Another cluster can be seen on the right side of the diagram: Companies with more than 2,000 tons per year are showing less than 0.5 kWh/l specific energy turnover. In general, we find a strong decreasing function indicated by the given interpolation line.

Lobbyists of regionalism promote the idea of changing this digression into a progression by adding the energy efforts for the global transport. This is proved in Figure 11. Here, all the energy terms of transportation and distribution are added to the data set of Figure 10. The cluster of small companies increases to an energetic performance from 1.5 up to 3.2 kWh per liter.

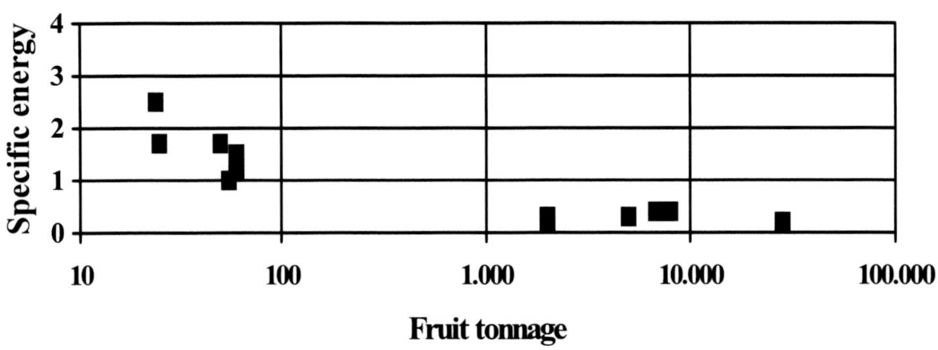

Figure 10. Specific energy turnover in kWh/l versus fruit tonnage in tons/a: production only.

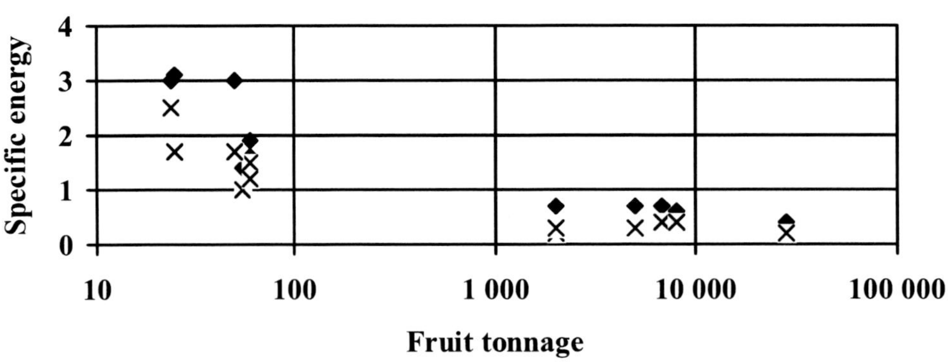

Figure 11. Specific energy turnover in kWh/l versus fruit tonnage in tons/a - production (x) plus transports and distribution (♦).

At the same time though the cluster of bigger companies increases only to values from 0.4 to 0.7 kWh per liter. The interpolation of these entire results gives again a strong declining function [Schlich 2000].

The assumption of changing the advantage of global business into a disadvantage because of the global transport energy is clearly falsified. Transportation and distribution efforts of small companies can be even higher than those for global food transportations. As clearly demonstrated small fruit companies need approximately 0.5 – 0.8 kWh/l for transportation and distribution, but surprisingly the global transportation takes just 0.1 – 0.3 kWh/l despite immense differences of the distance. Based upon primary data of all global and regional process steps it can be concluded that the specific energy turnover of fruit juices produced from small companies turns out to be higher than comparable fruit juices from companies with more than 2,000 tons per year, in all investigated cases.

Again, the theory of Ecology of Scale turns out to be appropriate as to describe the relation between specific energy turnover and business size. The pure distance between production and point-of-sale does not give any reliable information about the energetic backpack of the food. Global or European-continental juices are not to blame because of their supposed energy turnover. And, juices originating in the area close to the point-of-sale can compete in terms of energy just in case the business units are big enough to avoid energetic disadvantages.

Lamb Meat

The Germans on average consume about 0.7 kg of lamb and goat meat per capita and year. Altogether the total amount of this market is about 100,000 tons per year. Approximately 55 % of the yearly sale of such meat in Germany is imported, most of which coming from New Zealand. The New Zealand lamb meat is shipped as frozen food to Germany, transshipped at the seaport of Hamburg. Just about 45 % of the consumed lamb meat derives from regional origin of the German countryside. In view of these numbers it is clear that the German local or regional production of lamb meat cannot meet the entire demand of the consumers. Obviously, Germany as industrial country is reliant upon considerable import of that kind of meat products.

In order to research the hypothesis of "Ecology of Scale" local and regional German farmers are compared with the New Zealand process chain. The data research of the New Zealand process starts at the place of origin by personal evaluations of farms, slaughtering, dissectioning and freezing companies and shippers. Also here, all singular transportations from the farms to the seaport of Lyttleton (South Island of New Zealand) are included. The shipment from there to Hamburg takes 30 days. The lamb meat is deeply frozen at –30 °C and stored into reefers as standardized containers. The reefers are switched to cooling air on board of the vessel. The cooling system on board is energized by gas providing a compressor drive.

The vessel carries 1,403 reefers loaded with different goods, mostly lamb and cattle meat, furs and wool, all natural products of New Zealand. 97 reefers are bound for Hamburg, carrying 2,066,100 kg lamb meat ready-to-cook (net weight). The container vessel leaves Lyttleton eastward bound for Hamburg, usually via Port Chalmers (North Island of New Zealand), Santos (Brazil), Lisbon (Portugal), Zeebrugge (Belgium) and Tilbury (Great Britain). At each port some containers are unloaded, some others loaded. After reaching the final destination Hamburg the vessel returns immediately back home to New Zealand - via LeHavre (France), Dakar (Senegal), Cape Town (South Africa), Perth and Sydney (Australia) - fully loaded with European export goods. After a 2 months time period the entire crew changes, and the next trip around the world starts immediately.

By getting all the data of loading and unloading the vessel including all timetables, unload and load lists and the refuelling of the vessel it is possible to allocate the entire energy turnover of the overseas transportation and cooling efforts to the functional units [Fleissner 2001].

Before loading and after unloading at the seaports the reefers are stored at the seaport. To guarantee a continuous cooling the reefers are equipped with standardized backpacking cooling systems, using a 400 V power supply. Similar systems are used on the prime mover or HGV as to continue the cooling process up to the point of sale wherever that is. All components of this transportation and cooling technology are established and standardized worldwide to ensure an uninterrupted cooling process. To keep in memory, the seaports do not need any cooling storage facilities for the New Zealand lamb meat, due to an existing backpacking technology for the reefers.

To compare regional with global organized distributions we investigate local German farmers around the countryside, again, by personal interviews of farmers and shepherds, visiting places of breed, slaughter and dissectioning as well as researching the structure of regional marketing. Only with this intensive research it is possible to present valid data for the assessment of the specific energy turnover of regional process chains.

In contrast to the New Zealand breeding process, shepherds are necessary in Germany. Here, the sheep are guarded all day long and kept in fences during the night. Usually, German sheep are kept inside the stables for five months of the year, providing the herd with water and feed. This intensive proceeding of lamb meat production - with energetically expenses for daily shepherds, stables and additional feed in the wintertime - is only necessary in Germany, but not in New Zealand. One reason for that can be found in climatically differences between Germany and New Zealand. Secondly, the small density of population in New Zealand and the open countryside allow easy breeding of lamb in that country down under. In Germany, the shepherds usually take their car, to go to work. The additional efforts of daily shepherding in Germany and fencing in at night must be taken into account and added up to the specific energy turnover.

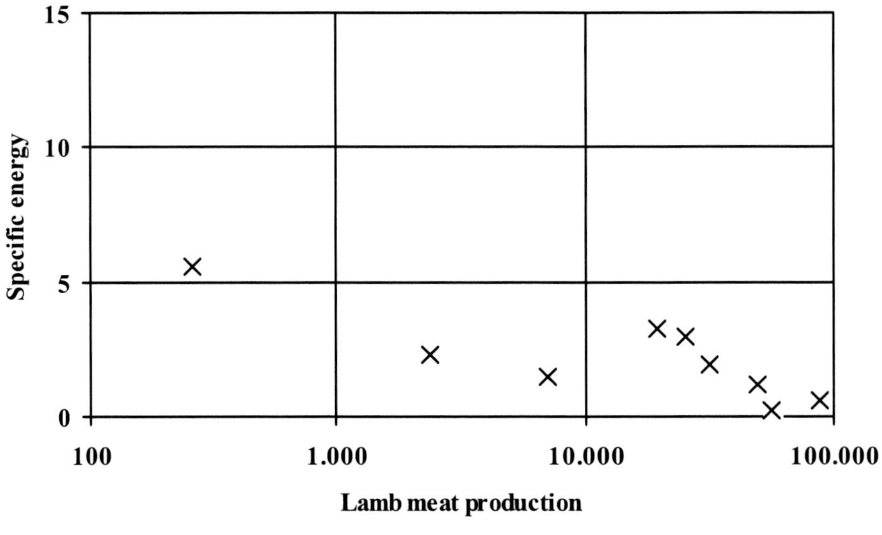

Figure 12. Specific energy turnover in kWh/kg versus lamb meat production in kg/a - production only.

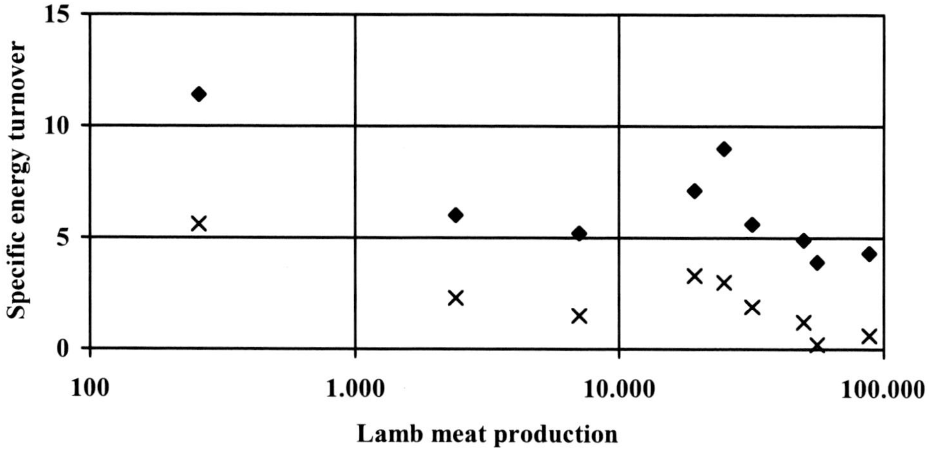

Figure 13. Specific energy turnover in kWh/kg versus lamb meat production in kg/a - production (x) plus transports and distribution (♦).

The reported data are researched again for one year of production and allocated to the functional unit of 1 kg lamb meat ready-to-cook. Thereby we get the yearly specific energy turnover in terms of kWh per kg. Figure 12 shows the results for production units dependent of the production throughput per year. Figure 13 adds the energy turnover of all the transports including the distribution to the point of sale.

Once again we find a declining logarithmic function indicated by the interpolation line. It is remarkable that small farmers require more specific energy than bigger units. The New Zealand farmers are able to produce lamb meat with only small expenses of energy. In figure 13 the energy expenses for transportation are added. Altogether, the global as well as the local transportation and distribution efforts are considered here.

The researched data of lamb meat demonstrate clearly two facts: first of all, the specific energy turnover is again dependent on the business size, as seen before with the data of fruit juices. Secondly, the sea transportation under strict cooling conditions in accordance to HACCP takes less energy than local transportation and distribution efforts.

The decrease is more weakly compared to fruit juices. This is to understand, because lamb meat cannot be concentrated as fruit juices before shipping. And, the transport temperature here is 30 °C meeting deep frozen requirements. On the other hand, the container vessel observed in 1999 carried just 1403 container. Today, much greater ships are in operation carrying some 8 000 container or even more. Hence, the specific energy turnover for these global operating container shipping should be even lower than recorded here. Just like beef from Argentina, no air cargo of lamb meat from New Zealand takes place. All meat products originating from New Zealand bound for Europe are shipped by container cargo via HGV and vessel, under strictly cooling conditions.

Pork

The most considerable meat products in Germany are made of pork. The trade and marketing of pork meat ready-to-cook achieves 39.2 kg per capita and year. This is about 65 % of the entire consumption of meat products in Germany. In sum, 4,500,000 tons are slaughtered per year, which is equal to 54.5 kg per capita and year. After removing bones, head, feet and some other parts which cannot be used for human food, 72 % (this is 39.2 kg per capita and year named above) of that amount is sold at the butcher´s.

The import of pork meat achieves about 1,500,000 tons per year. Regarding the German export of 1,200,000 tons per year, we have a self-supply of 94 %. As a matter of fact, Germany only imports pork meat from European partners, most of deriving from Denmark, the Netherlands, Poland and Hungary. Global import from countries outside EU does not play any significant role in this business. As to compare local, regional and European origin, we have research data from German and Hungarian farmers.

The pig fattening takes place in stables under use of power feed. In this special business we find an intensive farming, with considerable yearly turnover of animals. The fattening farms get the porklings from breeding specialists, as to bring them up until the pig reach the desired weight ready to slaughter. All conditions of breeding and fattening are highly standardized in order to achieve reproducible quality of the pork. The single steps of feeding, transporting and slaughtering are managed in accordance to the European food law, especially under strictly HACCP-conditions.

The described conditions of pork production are similar and comparable both in German regions and in Hungary. The process chains differ just regarding the question of business size. Small local farmers transport their pig production by pickup and trailer to the local slaughter, as in cases of beef and lamb meat production. The local process chain is similar to beef production described in Figure 7. The Hungarian process chain starts at the place of breeding porklings, which are transported by trucks to the fattening farm close by. After reaching the desired weight ready to slaughter the pigs are transported by HGV to the slaughterhouse. Both transports are getting back empty. The pork ready-to-cook is transported by international cooling transports, which connect Hungary to meat wholesaler in Germany.

All single steps of farming and transporting are considered, in terms of energy turnover, and then allocated to the functional unit of 1 kg pork ready-to-cook. The result of the pork investigation is shown in Figure 14, where the specific energy turnover is drawn over the business size.

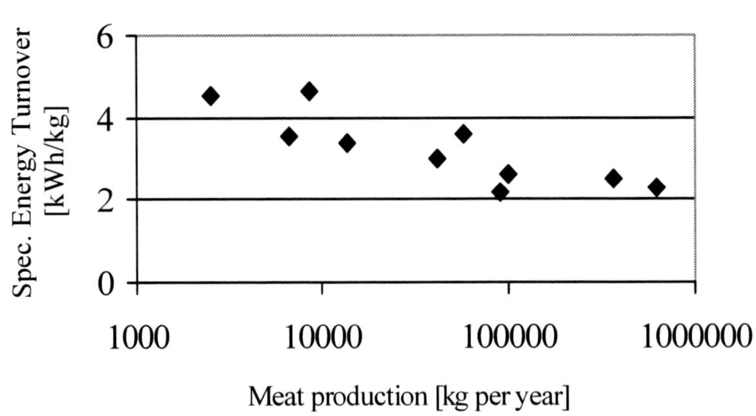

Figure 14. Specific energy turnover of pork – farming plus transport and distribution.

Obviously, there is an impact of the business size to observe. Again, the researched data support the idea of "Ecology of Scale" in the food business. Not the distance of food transport itself, but the business size determines the specific energy turnover of pork, similar to the other case studies. As a matter of fact, the efficiency in terms of energy increases with the size of the breeding and fattening farms. In addition, bigger transport units consume less energy per load and per mile, as assumed.

Hence, the transport distance itself does not play a main role regarding the question of energy turnover per kg pork. Furthermore, the data prove that German local or regional farmers can obtain energy efficiency, in case the farm is big enough. In this research there are some cases to observe, demonstrating the competitiveness of German farmers benchmarked to international process chains on the pork market, in terms of energy.

Wine

The consumption of wine in Germany is about 20 000 000 hectoliter per year. This corresponds to 24 Liter per capita and year. Germany is dependent on wine import because the national production of wine can quantitatively meet only 33 % of the consumption. This data does of course not consider any qualitative aspects of wine.

The import of bottled and non bottled wine is mainly originating from France, Italy and Spain. This imported amount is defined here as European-continental food. Global wines are originating from the Americas (California, Chile, and Argentina), from down under (Australia, and New Zealand) and from South Africa. Especially the import from South Africa enhances to about 1 % of the yearly consumption in Germany, with increasing tendency.

In order to investigate the hypothesis of "Ecology of Scale", four German regions (Baden, Hessische Bergstraße, Mosel, and Württemberg), two European-continental suppliers (France, Hungary) and one product of global origin (South Africa) are researched. In all cases wine bottled at the place of origin is involved. Hence, the heavy weight of the glass bottle has to be transported over global distances as well. The functional unit in all cases is here defined as hectoliter per year.

At all empiric investigations the area of the vineyard, the produced amount of wine, and all energy turnover are questioned. In addition, all steps of transport and distribution up to the point of sale are evaluated. This field research is very time expensive, because it has to be undertaken by personal interview of the process managers at the place of the process unit.

The result of all investigated process chains is given in Figure 15. Here the specific energy turnover in kWh per hl is presented over the yearly production in hl per year.

In sum, the specific energy turnover again decreases with the increasing size of the yearly production, as predicted by the theory of "Ecology of Scale". Nearly all data fit into the correlation. A production more than 1,000 hl per year is competitive in terms of energy, independent of the origin. A specific energy turnover of less about 100 kWh per hectoliter can be obtained, if the unit gets the required size.

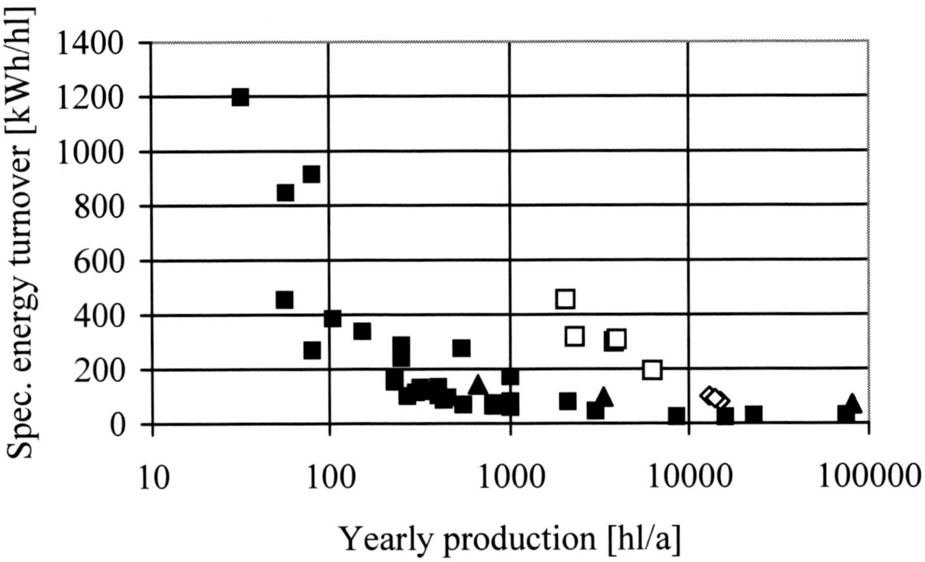

Figure 15. The specific energy turnover of wine from local, European-continental and global origin.

Two particular features can be observed. At first there is one German cooperative operating in a region as merger of more than 500 vineyards. The data coordinates in Figure 14 are marked by a white rhombus (◊). The yearly production achieves about 14,000 hl with a specific energy turnover of about 100 kWh per hl. A cooperative like this turns out to be an excellent opportunity as to achieve sufficient process size, especially regarding transport and distribution logistics. In other words: If any small wine-grower markets his production himself, the specific energy turnover can dramatically increase. Therefore, the foundation of a cooperative is recommended and supported by the actual data research.

Secondly, some special aspects of South-African vineyards are to observe. On the one hand, there are results within the digression, marked by a black triangle (▲). On the other hand, some other South-African units show up with results obviously above the digression. These producers are marked in Figure 15 by a white square (□). All wine cellars investigated in South-Africa operate storage facilities for red wine mash, which are cooled to 16 – 18 °C in order to improve the wine quality. But some of the cellars operate the storage facilities in open areas, under direct sunshine (□). These cellars definitely decline their energy balance, which can be seen in Figure 15 (□). The energy turnover decreases to reasonable values, if the cooling facilities are energy saving by insulation or operation in the basement (▲). Obviously, the data investigated here can be used to improve the operation in terms of energy turnover, by means of benchmarking and counseling.

After filling into bottles, the wine is transported by container transport to the seaport of Cape Town, and then by container vessel to Europoort Rotterdam in the Netherlands. The container is transloaded to an inland navigation vessel, bound for Mainz, Germany, to the place of wholesale. The last step to the point of sale is done by trucks. The specific energy turnover for all these steps of global transport is rather small. The data show, that the specific energy turnover of local or regional transports and distribution of wine are of similar dimension. That might be surprising at first sight. But, the efficiency of the international transports is quite high, due to the high load rate of the vessels and the perfect standardization of the containers.

Wine as a case study again gives evidence to the theory of Ecology of Scale. The specific energy turnover declines with increasing business size. The efforts for transport and distribution are of similar dimension, both in regional and in global process chains. Small units have to bear fixed costs in terms of energy. The fixed energy turnover has to be allocated to the small number of produced items. This effect burdens the energy balance for small production units.

Wine producers of local or regional kind are advised to merge into a foundation of regional cooperatives by establishing joint ventures, in order to increase the usage and fill rates of machinery and trucks. German regions are able to compete in terms of energy if the size of the business unit is sufficient. The case study finds a minimum business size in the range of 1,000 hl annual production, which is equal to about 125,000 bottles of wine (0.75 l each) per year. This number characterizes just a wine cellar of middle size in Germany, and not a big player in the wine market. Hence, the recent investigation and the data assessment lead to a minimum business size which can reasonably be managed, leading to sustainable structures in the local or regional area.

Conclusions and Outlook

All the presented case studies in the area of food business give evidence to the theory of Ecology of Scale. The predicted decrease of the specific energy turnover definitely correlates to the business size. The pure distance between place of origin and point of sale as such is rather insignificant, because the means of transportation and distribution have to be taken into account.

Global carriers use highly sophisticated process chains, which are based on reefers or standard containers shipped by vessel and HGV. By means of this technology the energy turnover to bridge global distances is rather small. In all cases the total energy turnover of transportation has to be allocated to the total amount of transported goods.

Local or regional carriers have to bridge smaller distances, but they use trucks or pickup cars and trailers with much smaller capacity of load. The local farmer delivers just one cattle to the butcher, because the local butcher does not need any more per week. Hence, the specific energy turnover increases dramatically, in particular with regard to the missing backload in the local or regional trucks. Altogether, the specific energy efforts for local, regional, European-continental or even global transports in many cases are of similar dimension. Greater business size is benefitted in terms of energy.

At first sight, this may be surprising, especially for agrarian lobbyists or local marketing managers. But, as a matter of fact, natural processes are quite similar. Scale related facts are well-known in nature. Considering for example the heat exchange of mammals to the environment, the specific surface of the body is the dominant factor. In absolute numbers big mammals of course are heavier than small mammals. But, in relative numbers, small mammals have a much higher specific surface compared with their bigger relatives. The specific surface is calculated by dividing the absolute surface by the body mass.

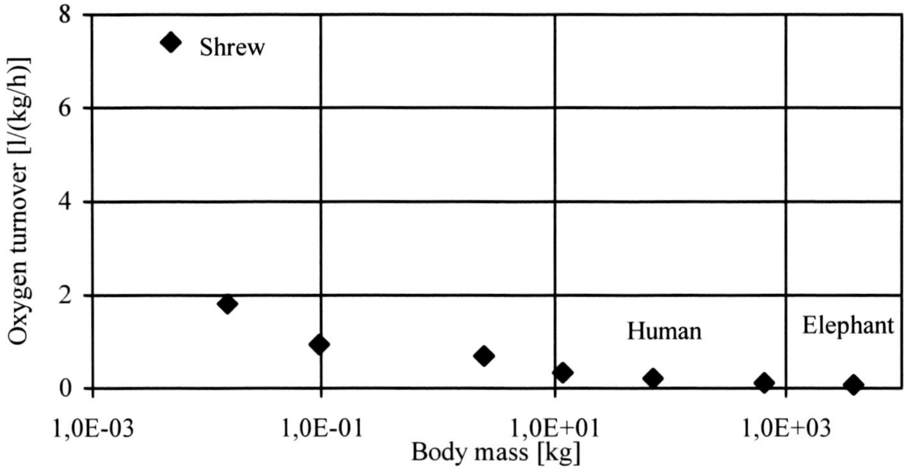

Figure 16. Specific oxygen turnover and body mass of different mammals[17] [Schlich 2004, Schmidt-Nielsen 1999].

The smaller the animal is, the higher is its specific surface, which is exchanging heat to the environment. The heat exchange of the body mass with the environment is proportional to the body surface, due to the thermodynamic laws of heat transfer. Because the temperature of mammals is usually higher than the temperature of their environment, a positive heat exchange rate from the body to the environment takes place. In order to keep the body temperature physiologically high, energy turnover is necessary. All the mammals survive by their nutrition. The most important task of the feed or food is the energy supply in order to keep the mammal warm.

Physiologists describe the energy turnover of living organisms, using the term metabolic rate. The metabolic rate (here: the specific oxygen turnover) per body mass of a small mammal is higher than the metabolic rate of a larger mammal (see Figure 16). The oxygen turnover of a shrew (body mass 0.0048 kg) is 7.4 Liter per kg and per hour, of a human (body mass 70 kg) 0.21 l/(kg*h) and of an elephant (body mass 3 833 kg) 0.07 l/(kg*h).

It is clear, that the energy turnover of mammals depends on the specific surface of their body. Small mammals have a high surface per kg body mass. The body surface area determines the heat exchange ratio. Hence the small mammals show up with a high metabolic rate, measured in terms of oxygen turnover per kg body mass and per hour.

The same correlation can be seen in Figure 17. Here the heart frequency in beats per minute is related to the body mass of mammals. Five typical examples are shown: The smallest mammal is the shrew (body mass = 4.8 g). Here we notice a heart frequency of 600 beats per minute. The heart of a little dog (body mass = 10 kg) beats about 100 times per minute. The pulse of a healthy human (body mass = 70 kg) beats approximately 60 times per minute. The horse (body mass = 500 kg) has 35 heart beats, and the elephant (body mass = 3 800 kg) only 25 heart beats per minute.

[17] The units are written in German notation. The number 1,0 is equal to 1.0 in American notation.

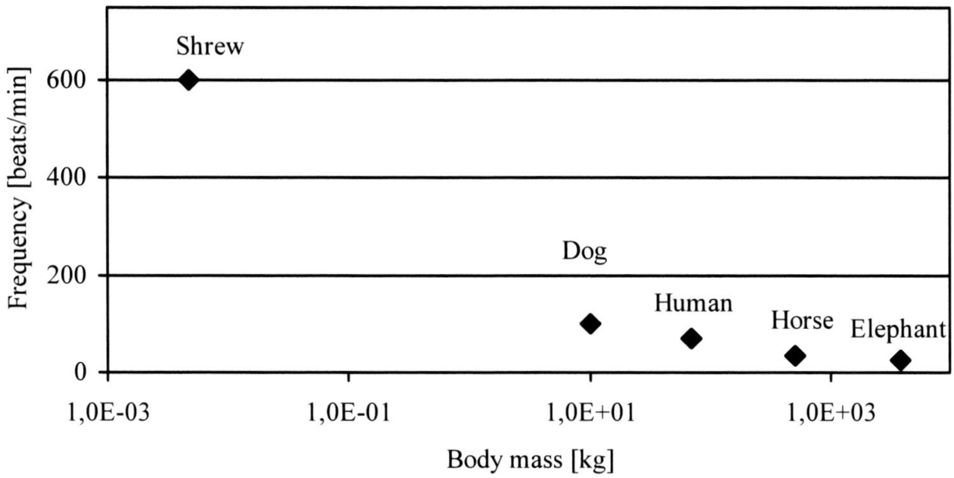

Figure 17. Heart frequency and body mass of different mammals [Schlich 2004, Stahl 1967].

All these data are related to the fact that the specific energy turnover of mammals depends upon their body size. Here we can see a strong declining correlation. The frequency of heart beats and the specific energy turnover correlate in reverse with the body mass. So the physiological data of mammals support the idea of Ecology of Scale. Bigger units have a smaller energy turnover, if the energy turnover is allocated to the body mass.

In consideration of these natural conditions of life, the idea of Ecology of Scale should be easier to understand. Further clues can be found at traffic carriers, or at domestic energy use. If we for example compare private cars with trains and planes, we find typical results, calculated in liters of gas per 100 Pkm[18]. A private car with 4 seats consumes 6.4 l per 100 Pkm; the plane with 200 seats takes 4.5 l per 100 Pkm and the train with 750 people on board just needs 2.2 l per 100 Pkm.

Again, we find in Figure 18 a declining function, demonstrating the disadvantage of smaller carriers.

The example of traffic carriers demonstrates clearly, that larger units need less energy per person than smaller. The same tendency can be seen in the energy turnover of private households, which is increasing in industrial nations like Germany. The smaller the number of persons per household the higher is the energy turnover per person. Especially singles are wasting a lot of energy [Schlich 2004b].

Nature of course knows this context very well. Ants and bees construct big colonies, to save energy and to control the climate of their home. Sheep crowd together, in order to minimize their heat exchanging surface, especially in the wintertime with snow and storms. In all cases, we find a more or less strong declining relation between the specific energy turnover and the size, supporting the theory of Ecology of Scale.

For the human economy of food, traffic, and housing, the coincidence of economic and ecological facts is obvious. And we cannot overlook, that human hedonism and individualism in the industrial countries turns out to be an energy wasting lifestyle.

18 The unit Pkm is the direct product of distance and number of persons. 100 Pkm could for example mean the transport of 100 people over 1 km or the transport of 1 person over 100 km.

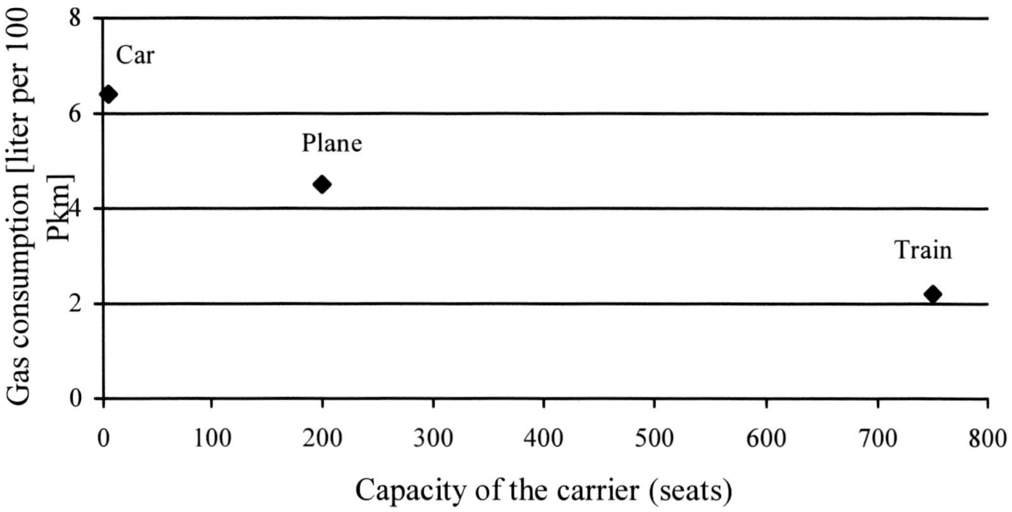

Figure 18. Specific energy of traffic carriers in litre per 100 Pkm over the capacity of the carrier.

Even the organic quality of food is dependent upon the efficiency and logistics of the entire process chain. Unfortunately, this matter plays no role in the business of labeling organic food in Germany, because labeling of organic food only intends to look after the conditions of farming.

All the conclusions are valid for the case studies investigated until today. This restriction is quite common in science. But at the same time some fundamental conclusions are obvious. As demonstrated very clearly, global process chains are not to blame in terms of energy turnover. Even local or regional process chains can waste energy if the business size is too small. On the other hand, the data allow the definition of minimum business size, which can be achieved in local or regional networks quite easily. Such examples can be seen in any case study performed. But the supposed advantage of regionalism in general is disproved by empiric investigation. And the results observed give evidence to a more or less natural principle called the Ecology of Scale.

Acknowledgement

The scientific research presented here has been kindly supported by Deutsche Forschungsgemeinschaft (DFG)[19], Deutscher Akademischer Austauschdienst (DAAD)[20], and Justus-Liebig-Universität Gießen.

Publishing this contribution the author is supported by quite a remarkable number of scientists, who presented their bachelor thesis, master thesis or dissertation at Justus-Liebig-University of Giessen, Germany. The research started in 1997 and is still going on. Without the ideas and the particular research of students, graduate and postgraduate scientist in local,

[19] Research Association of all German universities.

[20] German Academic Exchange Service.

regional, European-continental and global areas of food business we never would have any scientific progress.

Last but not least I would like to thank all our cooperation partners at companies and universities around the world, who help us to identify process chains, subsystems, energy turnover and business size. Thank you all for this essential support and your personal contributions to the present research.

References

[DIN EN ISO 14040] Normenausschuss Grundlagen des Umweltschutzes (NAGUS) im DIN Deutsches Institut für Normung e. V.: Produkt-Ökobilanz Prinzipien und allgemeine Anforderungen. Deutsche Fassung der ISO 14040 :1997.

[EG 178/2002] European Union: Verordnung 178 zur Festlegung der Allgemeinen Grundsätze und Anforderungen des Lebensmittelrechts, zur Errichtung der Europäischen Behörde für Lebensmittelsicherheit und zur Festlegung von Verfahren zur Lebensmittelsicherheit: http://www.bmelv.de (2002).

[EG 852/2004] European Union: Verordnung 852 über Lebensmittelhygiene: http://www.bmelv.de (2004).

[EG 853/2004] European Union: Verordnung 853 mit spezifischen Hygienevorschriften für Lebensmittel tierischen Ursprungs: http://www.bmelv.de (2004).

[EG 854/2004] European Union: Verordnung 854 mit besonderen Verfahrensvorschriften für die amtliche Überwachung von zum menschlichen Verzehr bestimmten Erzeugnissen tierischen Ursprungs: http://www.bmelv.de (2004).

[EWG 2092/91] European Union: EG-Öko-Verordnung: http://www.bmelv.de (1991).

[Fleissner 1999] Fleissner U, Schlich E (1999): Von der Orange zum Orangensaft - die energetische Betrachtung der Produktion eines Lebensmittels. In: Der Mathematische und Naturwissenschaftliche Unterricht. Organ des Deutschen Vereins zur Förderung des mathematischen und naturwissenschaftlichen Unterrichts e. V.. (52) 1999, *Heft* 2, S. 94 – 101, Ferdinand Dümmlers Verlag Bonn.

[Fleissner 2001] Energetische Bewertung der Bereitstellung ausgewählter regionaler und überregionaler Lebensmittel. Aachen: Shaker 2002. Zugleich: Dissertation an der Justus-Liebig-Universität Gießen (2001). ISBN 3-8265-9797-4.

[Jungbluth 1998] Jungbluth N: Ökologische Betrachtungen des Bedürfnisfeldes Ernährung. Arbeitsgruppen, Methoden, Erkenntnisse. Working Paper des Fachbereiches Umweltnatur- und Umweltsozialwissenschaften (UNS) der Eidgenössischen Technischen Hochschule Zürich (ETH). Zürich (1998).

[Jungbluth 2000] Jungbluth N: Umweltfolgen des Nahrungsmittelkonsums: Beurteilung von Produktmerkmalen auf Grundlage einer modularen Ökobilanz. Dissertation ETH 13499/2000. Verlag Ökoinstitut e.V. Freiburg (2000).

[Koerber 1999] Koerber K v, Kretschmer J: Der Anspruch auf Nachhaltigkeit im Ernährungsbereich. Wie zukunftsfähig ist unser Ernährungsstil? In: aid-Verbraucherdienst 44 -4/99, S. 88-95 (1999).

[Kranendonk 1993] Kranendonk S, Bringezu S: Major Material Flows Associated with Orange Juice Consumption in Germany. In: *Fresenius Envir Bull* 2:455-460. (1993).

[Schlich 2000] Schlich E, Fleissner U (2000): Energetischer Vergleich der Produktion und Distribution ausgewählter regionaler und globaler Lebensmittel. In: Wissenschaft und Praxis mit regionaler und globaler Bedeutung. Dokumentation der 1. Hochschultagung des Fachbereichs Agrarwissenschaften, Ökotrophologie und Umweltmanagement. Fachverlag Köhler, Gießen 2000, ISBN 3-934229-72-7, 113 – 121.

[Schlich 2004] Schlich E, Schlich M: The Ecology of Scale: Further Examples and Comments. InLCA 2004.

[Schmidt-Nielsen 1999] Schmidt-Nielsen K: Physiologie der Tiere. Heidelberg (1999).

[StaBu 2007] Statistisches Bundesamt: http://www.destatis.de (2007).

[Stahl 1967] Stahl W R: Scaling of respiratory variables in mammals. *In: J. Appl. Physiol.* 22 (1967), 453-460.

[UBA 1992] Umweltbundesamt (Editor): Texte 38/92: Ökobilanzen und Produkte - Bedeutung - Sachstand - Perspektiven. Berlin (1992).

[Weizsäcker 1995] Weizsäcker E U v et al: Faktor 4: Doppelter Wohlstand - halbierter Naturverbrauch; der neue Bericht an den Club of Rome. München: Droemer Knaur (1995).

[Zamboni 1994] Zamboni M: Grobabschätzung des Energieaufwandes für die Bereitstellung von ausgewählten Getränken und Nahrungsmitteln. Im Auftrag von Greenpeace Schweiz und des Konsumentinnenforums Schweiz, Zürich (1994).

In: New Research on Energy Economics
Editors: P.G. Caldwell and E.V. Taylor, pp. 107-134
ISBN: 978-1-60456-354-2

Chapter 3

OIL PRICE EFFECTS AND ASIA PACIFIC EQUITY MARKETS

Perry Sadorsky[1]
Schulich School of Business, York University
4700 Keele Street, Toronto, Ontario M3J 1P3, Canada

Abstract

The Asia Pacific region of the world is home to some of the fastest growing economies in the global economy. The vast majority of countries in the Asia Pacific region are net importers of crude oil and rising oil prices could impact economic and financial activity. This study uses a multifactor market model to investigate the impact that oil price changes have on the equity markets of the Asia Pacific economies. Results are presented to show that oil price movements are an important factor influencing the stock returns for approximately half of the countries studied. Asymmetric oil price changes are important factors influencing stock markets for the countries of India, Japan, South Korea, and Sri Lanka. Asymmetric oil price volatility changes are important determinants of stock prices only in the case of Singapore. Managers, investors, and policy makers should find these results useful.

Keywords: Emerging market economies; stock market risk; oil price risk

1. Introduction

The Asia Pacific region of the world is home to some of the fastest growing economies in the global economy. Between 1993 and 2003, for example, real GDP grew on average by 8.9% per year in China and 6.2% per year in India (The Economist (2006)). South Korea, Malaysia and Singapore each had economies that grew at an annual average rate of 5.4%, 5.2% and 5.2% respectively over the period 1993 to 2003. In comparison, Japan's economy grew by an average rate of 1.3% per year over the same time period while the global economy grew by an annual average of 3.6%. As has typically been the case for rapidly growing economies, economic growth leads to an increased demand for the factors of production like capital, labour, energy (including oil and oil related products), and materials. For the fast growing economies of the Asia Pacific region, the increase in oil consumption has been dramatic.

Currently, the Asia Pacific region of the world accounts for 29.1 % of world oil consumption and this is second only to North America, which accounts for 29.5% of world oil consumption (Table 1). Between 1995 and 2005, oil consumption in the Asia Pacific region grew by 32.5% which makes it the region with the second greatest increase in oil consumption over this time period. Between 1995 and 2005, oil consumption in the world's two most populous countries, China and India grew by 105.9% and 57.2% respectively. By comparison, oil consumption in Japan and the Philippines over this same time period actually declined by 7.3% and 8.7% respectively. China has the potential to seriously impact global oil demand. If China increases oil consumption at a rate of 5% a year over the next 25 years then China's daily oil consumption in 2030 will be 23,664 thousand barrels of oil per day which is greater than daily oil consumption in the United States in 2005. The big consumers of oil are not, however, the countries with large proven oil reserves. At the end of 2005, the United States had just 2.4% of world proven oil reserves while China had 1.3% (BP (2006)).

The purpose of this present paper is to contribute to the literature on stock markets and energy prices by studying the impact of oil price effects on Asia Pacific stock market returns. The approach taken in this paper uses an international multifactor model that allows for both oil price changes and oil price volatility to impact stock returns. Results are presented showing the impact of both symmetric and asymmetric oil price changes (and oil price volatility changes) on equity returns.

This paper is organized as follows. Section 2 presents a brief review of the literature regarding the relationship between oil prices and stock prices. Section 3 presents and discusses the data and methodology. Section 4 reports on the empirical results. Section 5 concludes the paper.

2. Literature Review

The relationship between oil price changes and stock prices can be explained using an equity pricing model. In an equity pricing model, the price of equity at any point in time is equal to the expected present value of discounted future cash flows (Huang, Masulis and Stoll (1996)). Capital, labour, energy (including oil and oil related products), and materials are important inputs into the production of most goods and services and changes in the prices of these inputs affect cash flows. Rising oil prices, which, in the absence of complete substitution effects between the factors of production, increase the costs of producing goods and services. Higher costs of production dampen cash flows and reduce stock prices. Rising oil prices also impact the discount rate used in the equity pricing formula. Rising oil prices are often indicative of inflationary pressures which central banks can control by raising interest rates. Higher interest rates make bonds look more attractive relative to stocks leading to a fall in

[1] E-mail address: psadorsk@schulich.yorku.ca, Telephone: +1 416 736 5067 Fax: +1 416 736 5687.

stock prices. The overall impact of rising oil prices on stock prices depends on whether a company is a consumer or producer of oil and oil related products. Since there are more companies in the world that consume oil than produce oil, the overall global impact of rising oil prices on stock markets is expected to be negative.

Table 1. Oil consumption

Oil: Consumption* **Thousand barrels daily**	**1995**	**2000**	**2005**	**Change 2005 over 2004**	**2005 share of total**	**Change 2005 over 1995**
USA	17725	19701	20655	-0.2%	24.6%	16.5%
Total North America	21150	23522	24875	0.1%	29.5%	17.6%
Total S. and Cent. America	4133	4661	4776	2.8%	5.8%	15.5%
Total Europe and Eurasia	19703	19564	20350	0.9%	25.1%	3.3%
Total Middle East	4240	4735	5739	4.4%	7.1%	35.4%
Total Africa	2198	2458	2763	4.4%	3.4%	25.7%
Australia	781	837	884	2.7%	1.0%	13.1%
China	3395	4772	6988	2.9%	8.5%	105.9%
China Hong Kong SAR	198	201	285	-9.6%	0.4%	43.9%
India	1580	2254	2485	-3.5%	3.0%	57.2%
Indonesia	820	1049	1168	1.4%	1.4%	42.5%
Japan	5784	5577	5360	1.4%	6.4%	-7.3%
Malaysia	381	441	477	-3.3%	0.6%	25.2%
New Zealand	125	134	152	1.1%	0.2%	21.5%
Pakistan	315	373	353	8.8%	0.5%	12.0%
Philippines	344	348	314	-6.9%	0.4%	-8.7%
Singapore	617	654	826	11.0%	1.1%	34.0%
South Korea	2009	2229	2308	0.8%	2.7%	14.9%
Taiwan	713	816	884	0.2%	1.1%	23.9%
Thailand	717	725	946	4.0%	1.2%	32.1%
Other Asia Pacific	245	363	445	4.3%	0.5%	82.1%
Total Asia Pacific	18081	20839	23957	1.5%	29.1%	32.5%
TOTAL WORLD	69506	75779	82459	1.3%	100.0%	18.6%
Of which European Union 25#	13766	14402	14772	0.7%	18.3%	7.3%
OECD	44475	47646	49254	0.4%	59.2%	10.7%
Former Soviet Union	4410	3623	3936	1.9%	4.9%	-10.8%
Other EMEs	20621	24511	29270	2.6%	36.0%	41.9%

* Inland demand plus international aviation and marine bunkers and refinery fuel and loss.
BP Statistical Review of World Energy, June 2006 (www.BP.com).

To date, a number of published papers have investigated the relationship between oil price movements and stock prices in developed countries (see for example, Chen, Roll and Ross (1986), Hamao (1989), Kaneko and Lee (1995), Ferson and Harvey (1995), Jones and Kaul (1996), Huang, Masulis and Stoll (1996), Sadorsky (1999, 2003), and Faff and Brailsford (1999)). Recent papers by Papapetrou (2001), Hammoudeh and Aleisa (2004), Hammoudeh and Choi (2006), Basher and Sadorsky (2006), and Nandha and Hammoudeh (2007) investigate the impact of oil price changes on stock price in emerging economies. The consensus from most of these papers is that oil price changes do have some impact on stock prices.

Chen, Roll and Ross (1986) find that interest rates, inflation rates, bond yield spreads, and industrial production have risk that is priced in the stock market but they did not find any evidence that oil price risk is rewarded by the stock market. Hamao (1989), using the approach of Chen, Roll and Ross (1986), finds no evidence for the pricing of an oil price factor in Japanese equity data. Kaneko and Lee (1995) use a more recent sample of Japanese equity data and do find some evidence in favour of an oil price factor impacting stock returns. Ferson and Harvey (1995) find evidence that an oil price risk factor does have a statistically significant but different impact on the eighteen equity markets that they study. Jones and Kaul (1996) find that the reaction of Canadian and U.S. stock prices to oil price shocks can be completely accounted for by the impact of oil price shocks on real cash flows. The results for Japan and the United Kingdom are, however, not as strong. Huang, Masulis and Stoll (1996) find that oil futures returns do lead some individual oil company stock returns but oil futures returns do not have much impact on broad based market indices like the SandP 500. Sadorsky (1999) finds that oil price changes and oil price volatility both play important roles in affecting real US stock returns and that oil price changes have asymmetrical impacts on stock prices. Increases in oil prices have a greater impact on stock prices than do decreases in oil prices. Faff and Brailsford (1999) find that oil price changes do impact the stock price returns of some sectors (like oil and gas, paper and packaging, and transportation) of the Australian economy. Papapetrou (2001) finds that changes in oil prices influence real economic activity, stock prices and employment in Greece. Hammoudeh and Aleisa (2004) find that the Saudi Arabian stock market has a bi-directional relationship between oil prices and stock prices. Basher and Sadorsky (2006) find strong evidence that oil price risk impacts stock price returns in emerging markets although the exact relationship depends somewhat on the data frequency being used. The conditional relationship is not, however, symmetrical. Hammoudeh and Choi (2006) study stock markets in the Gulf Cooperation Council (GCC) and find that the U.S. T bill rate has a direct effect on these stock markets while oil prices and the SandP 500 each have indirect effects. In a study of 15 countries in the Asia Pacific region, Nandha and Hammoudeh (2007) find that oil prices do impact the stock markets in the Philippines and South Korea. This present paper studies the same countries studied by Nandha and Hammoudeh (2007) and it is thus important to point out the differences between this present paper and theirs. Nandha and Hammoudeh (2007) use weekly data from May 4, 1994 to June 30, 2004 while in this present paper monthly data is used covering the much longer period December 1987 to December 2006. The Nandha and Hammoudeh (2007) methodology focuses more on the relationship between weekly returns and systematic risk (similar to arbitrage pricing theory which looks into whether or not the ex ante risk of oil shocks is priced in the market) while this present paper uses a multifactor model to investigate ex post sensitivity of stock returns to changes in oil prices and changes in oil price volatility. In addition, this present paper extends the existing literature on the relationship between oil prices and emerging market stock prices by investigating the relationship between oil price volatility and stock returns.

While changes in oil prices can impact stock prices, it is also possible for changes in oil price volatility to impact stock prices. Price return volatility can be an accurate measure of the rate of information flow in a financial market (Ross (1989). Evidence that volatility is correlated across markets would imply dependence in the information flow process. Oil prices respond to global oil demand and supply conditions, geopolitics, institutional arrangements (OPEC), and the dynamics of the futures market (Sadorsky (2004)). Unanticipated changes in any of these four factors can create volatility, and hence risk, in oil futures prices. Oil price volatility measures information flows from global oil demand and supply conditions, geopolitics, institutional arrangements stemming from OPEC's actions, and the dynamics of the futures market. To date, there is some evidence that oil price volatility impacts stock prices. Huang, Masulis and Stoll (1996) find that oil futures volatility leads the petroleum stock index volatility. Sadorsky (1999) finds that changes in oil price volatility do impact real U.S. stock returns. Sadorsky (2003) finds the changes in oil price volatility impacts technology stock price returns volatility.

While it is likely that oil price changes and/or oil price volatility have an impact on stock prices, it is also likely that at least for some countries the relationship is asymmetrical. A number of previous papers have observed and provided explanations for the fact that oil price increases seem to have a greater impact on economic activity than do oil price decreases (see for example, Mork (1989), Hooker (1996,2002), Hamilton and Herrera (2004), Davis and Haltiwanger (2001), Balke et al. (2002), Lee et al. (2001), Cunado and Perez de Garcia (2003)). In their survey of the literature on oil and the macroeconomy, Jones et al (2004) report that the asymmetric response of economic activity to oil price changes is a basic finding. Adjustment costs and financial stress are popular explanations for the asymmetry, but to date there is no widely agreed upon explanation. There has also been some research done looking into the impact that asymmetric oil price shocks have on stock prices. Sadorsky (1999) and Basher and Sadorsky (2006) find support for the hypothesis that oil price shocks asymmetrically affect stock prices.

3. Methodology and Data

The data for this present study consists of monthly closing prices on 15 stock markets in the Asia Pacific region of the world and the Morgan Stanley Capital International (MSCI) World Equity Market Index. All equity markets are measured using gross price indexes which include dividends. The use of monthly data avoids the noise associated with daily or weekly data and monthly data provide a direct link with other macroeconomic data. The equity market data are available from Datastream and, for most countries, cover the period December 1987 to December 2006 for a total of 229 monthly observations. Due to data availability, the data sets for China, India, Pakistan, and Sri Lanka are slightly shorter and begin in December 1992. All of the data are in U.S. dollars so that investment decisions are made from the perspective of a U.S. investor or an international investor who has a U.S. dollar trading account. The countries included in the study are, Australia, China, Hong Kong, India, Indonesia, Japan, Korea (South), Malaysia, New Zealand, Pakistan, Philippines, Singapore, Sri Lanka, Taiwan, and Thailand.

Monthly excess stock returns are calculated by subtracting the monthly yield on a three month U.S. Treasury Bill from the continuously compounded monthly stock market returns. The average monthly excess stock returns for a country are small in comparison to its standard deviation (Table 2). Hong Kong has the highest average excess monthly return (0.73 %) while China has the lowest average excess monthly return (-0.53 %). Only two (China and Japan) of the fifteen countries have negative average excess returns. The risk of Asia Pacific stock market returns is shown by the standard deviations. Indonesia has the highest standard deviation. The world market returns have the lowest standard deviation, illustrating the benefits of portfolio diversification. It is important to realize that the period of time under study, December 1987 to December 2006, was a very volatile period for stock markets around the world. Major events during this time period included the Asian financial crises in 1997 and 1998, Russia, Brazil and Long Term Capital Management close to default in 1998, Y2K scares in 1999, the bursting of the technology stock bubble in the spring of 2000, and the September 11, 2001 terrorist attacks on the World Trade Center in New York City. Most of the stock returns exhibit some form of either skewness or kurtosis. Unit root tests (Table 2c) confirm that each time series is stationary.

The multifactor market model is estimated by including risk factors for world stock market returns, interest rates, exchange rates, oil prices, and oil price volatility.

$$R_{it} = c + \beta_{mi} MR_t + \beta_{ri} Irate_t + \beta_{ei} EXCH_t + \beta_{oi} Oil_t + \beta_{ovi} Oilvol_t + \varepsilon_{it} \qquad (1)$$

In equation (1) the country excess returns are R_{it} (i=1,...,15 denotes the country and t denotes the time period), world market excess returns are MR_t, $Irate_t$ is the interest rate variable, exchange rate returns are $EXCH_t$, oil price returns are Oil_t, and oil price volatility is $Oilvol_t$. World market excess returns are measured by the excess returns on the MSCI World Market Index. Oil returns are measured as the log difference of the end of month closing prices on the West Texas Intermediate (WTI) crude oil futures contract which trades on the New York Mercantile Exchange (NYMEX).

Table 2a. Summary statistics for stock returns, oil price returns, oil volatility, and interest rates

	Australia	China	Hong Kong	India	Indonesia	Japan	Korea	Malaysia	New Zealand	Pakistan
Mean	0.63	-0.53	0.73	0.63	0.51	-0.29	0.28	0.26	0.22	0.39
Median	0.56	0.10	0.62	1.30	0.88	-0.48	-0.47	0.67	0.78	-0.26
Maximum	15.59	38.02	28.35	19.58	65.79	21.10	52.98	40.14	23.73	31.68
Minimum	-16.87	-32.01	-34.48	-19.74	-52.43	-22.23	-37.90	-36.39	-22.78	-47.79
Std. Dev.	5.27	10.72	7.55	8.22	14.24	6.52	10.91	8.83	6.55	11.16
Skewness	-0.26	0.18	-0.22	-0.20	0.35	0.05	0.28	-0.24	-0.17	-0.30
Kurtosis	3.43	4.44	5.51	2.56	7.34	3.55	6.08	6.97	4.03	5.33
Jarque-Bera	4.27	15.54	61.66	2.45	183.68	3.02	92.90	151.62	11.24	40.64
Probability	0.12	0.00	0.00	0.29	0.00	0.22	0.00	0.00	0.00	0.00
Observations	228	168	228	168	228	228	228	228	228	168

	Philippines	Singapore	Sri Lanka	Taiwan	Thailand	World	Oil price	Oil vol	Twex	Irate
Mean	0.17	0.45	0.27	0.19	0.14	0.36	0.57	9.76	0.26	-0.01
Median	0.33	0.83	-0.06	0.38	0.64	0.79	0.86	9.10	0.28	0.00
Maximum	35.70	22.55	39.73	37.88	35.56	10.04	36.89	50.25	3.85	0.45
Minimum	-34.99	-23.08	-29.39	-41.69	-42.00	-14.72	-27.79	3.01	-3.31	-0.79
Std. Dev.	9.43	6.98	9.96	11.01	11.62	4.02	9.04	4.80	1.18	0.13
Skewness	-0.08	-0.55	0.35	-0.08	-0.43	-0.59	0.17	3.56	0.14	-0.90
Kurtosis	4.79	5.47	5.11	4.38	4.92	3.87	4.52	25.96	3.49	8.60
Jarque-Bera	30.70	69.40	34.62	18.45	42.07	20.36	23.11	5489.54	3.05	328.86
Probability	0.00	0.00	0.00	0.00	0.00	0.00	0.00	0.00	0.22	0.00
Observations	228	228	168	228	228	228	228	228	228	228

Monthly returns are shown for each variable except Irate which is the first difference in the TED spread.

Table 2b. Summary statistics for foreign exchange rates

	Australia	China	Hong Kong	India	Indonesia	Japan	Korea	Malaysia	New Zealand	Pakistan
Mean	-0.04	0.32	0.00	0.54	0.75	-0.04	0.06	0.15	-0.03	0.55
Median	-0.09	0.00	0.00	0.14	0.35	0.13	0.00	0.00	0.02	0.14
Maximum	5.92	40.44	0.35	19.56	67.72	8.06	36.88	15.12	6.52	8.14
Minimum	-5.86	-1.53	-0.59	-5.91	-26.88	-10.52	-8.93	-14.48	-7.41	-2.41
Std. Dev.	2.24	2.95	0.10	1.88	6.99	2.67	3.22	2.16	2.32	1.32
Skewness	0.14	11.69	-1.32	5.15	4.32	-0.52	6.75	1.07	0.15	2.86
Kurtosis	2.81	153.79	13.32	50.35	43.68	4.13	77.74	28.25	3.37	13.43
Jarque-Bera	1.07	221189.30	1077.92	22305.89	16433.72	22.56	54794.46	6099.85	2.15	1343.65
Probability	0.59	0.00	0.00	0.00	0.00	0.00	0.00	0.00	0.34	0.00
Observations	228	228	228	228	228	228	228	228	228	228

	Philippines	Singapore	Sri Lanka	Taiwan	Thailand
Mean	0.38	-0.12	0.55	0.05	0.15
Median	0.12	-0.14	0.34	-0.01	-0.03
Maximum	13.78	5.64	9.32	6.83	21.02
Minimum	-7.11	-5.70	-5.72	-4.58	-14.16
Std. Dev.	2.13	1.29	1.10	1.32	2.85
Skewness	1.68	0.18	1.93	0.49	2.57
Kurtosis	12.60	6.80	24.17	7.43	25.84
Jarque-Bera	982.95	138.48	4398.76	195.24	5206.42
Probability	0.00	0.00	0.00	0.00	0.00
Observations	228	228	228	228	228

Monthly returns are shown for each variable.

Table 2c. Unit root tests for stock returns, oil price returns, oil volatility, and interest rates

	Australia	China	Hong Kong	India	Indonesia	Japan	Korea	Malaysia	New Zealand	Pakistan
ADF										
T test	-16.31	-11.61	-14.24	-11.93	-12.67	-14.45	-14.49	-7.73	-16.02	-12.69
P value	0.00	0.00	0.00	0.00	0.00	0.00	0.00	0.00	0.00	0.00
lag	0.00	0.00	0.00	0.00	0.00	0.00	0.00	1.00	0.00	0.00
PP										
T test	-16.42	-11.55	-14.22	-11.93	-12.61	-14.45	-14.50	-12.71	-16.02	-12.70
P value	0.00	0.00	0.00	0.00	0.00	0.00	0.00	0.00	0.00	0.00
lag	8.00	10.00	10.00	1.00	4.00	1.00	3.00	1.00	3.00	2.00
KPSS										
LM test	0.23	0.39	0.13	0.35	0.19	0.12	0.14	0.10	0.29	0.29
lag	12.00	8.00	9.00	0.00	5.00	2.00	1.00	4.00	2.00	2.00

	Philippines	Singapore	Sri Lanka	Taiwan	Thailand	World	Oil price	Oil vol	Twe	Irate
ADF										
T test	-11.98	-14.64	-12.09	-13.52	-8.85	-15.23	-13.81	-4.45	-10.07	-8.29
P value	0.00	0.00	0.00	0.00	0.00	0.00	0.00	0.00	0.00	0.00
lag	0.00	0.00	0.00	0.00	1.00	0.00	0.00	2.00	0.00	5.00
PP										
T test	-11.97	-14.64	-12.12	-13.54	-14.51	-15.25	-13.92	-10.20	-9.69	-21.55
P value	0.00	0.00	0.00	0.00	0.00	0.00	0.00	0.00	0.00	0.00
lag	1.00	0.00	3.00	2.00	6.00	7.00	15.00	7.00	9.00	44.00
KPSS										
LM test	0.21	0.14	0.28	0.07	0.20	0.09	0.12	0.10	1.57	0.43
lag	3.00	0.00	3.00	4.00	5.00	7.00	13.00	9.00	0.00	22.00

The lag length for the ADF test regression is set using the Schwarz information criterion. The bandwidth for the PP and KPSS test regressions are set using a Bartlett kernel.

Table 2d. Unit root tests for foreign exchange rates

	Australia	China	Hong Kong	India	Indonesia	Japan	Korea	Malaysia	New Zealand	Pakistan
ADF										
T test	-10.51	-13.81	-13.95	-11.82	-11.57	-10.94	-10.57	-11.67	-11.03	-9.58
P value	0.00	0.00	0.00	0.00	0.00	0.00	0.00	0.00	0.00	0.00
lag	1.00	0.00	0.00	0.00	0.00	0.00	1.00	0.00	0.00	1.00
PP										
T test	-11.20	-13.81	-13.92	-11.90	-11.67	-10.90	-8.05	-11.69	-10.97	-9.75
P value	0.00	0.00	0.00	0.00	0.00	0.00	0.00	0.00	0.00	0.00
lag	3.00	2.00	5.00	4.00	9.00	4.00	21.00	3.00	2.00	2.00
KPSS										
LM test	0.17	0.43	0.06	1.00	0.13	0.08	0.13	0.10	0.23	0.48
lag	3.00	1.00	3.00	6.00	0.00	0.00	4.00	5.00	5.00	5.00

	Philippines	Singapore	Sri Lanka	Taiwan	Thailand
ADF					
T test	-9.80	-11.35	-11.33	-10.85	-11.28
	Philippines	Singapore	Sri Lanka	Taiwan	Thailand
P value	0.00	0.00	0.00	0.00	0.00
lag	0.00	0.00	0.00	0.00	0.00
PP					
T test	-9.77	-11.27	-11.26	-10.86	-11.24
P value	0.00	0.00	0.00	0.00	0.00
lag	3.00	13.00	2.00	8.00	8.00
KPSS					
LM test	0.11	0.29	0.19	0.14	0.13
lag	7.00	6.00	4.00	2.00	1.00

The lag length for the ADF test regression is set using the Schwarz information criterion. The bandwidth for the PP and KPSS test regressions are set using a Bartlett kernel.

The WTI futures contract is the most widely traded oil futures contract in the world and is used as a benchmark to set other oil product related prices. Moreover, The NYMEX oil futures contract is the most heavily traded futures contract on a physical commodity in the world and therefore represents an efficient flow of information between buyers and sellers (www.nymex.com/lsco_pre_agree.aspx). Monthly oil price volatility is measured using the square root of the sum of squared daily returns over a month. The oil price futures series is available from Datastream.

Extending beta pricing models to an international setting requires a number of assumptions including integrated capital markets, purchasing power parity, and no informational or transactions costs or taxes. When purchasing power parity does not hold then individuals face exchange rate risk when investing internationally (Adler and Dumas (1983)). Exchange rate risk can be separated into an individual factor for each country or it can be approximated using a single variable. In this study we follow both approaches. In the first set of results, individual country exchange rates via the U.S. dollar (FX) are used as the exchange rate variable. In the second set of results, we follow Ferson and Harvey (1994) and use a single variable to approximate exchange rate risk. The variable Twex is the log difference of a weighted average of the foreign exchange value of the U.S. dollar against a subset of the broad index currencies that circulate widely outside the country of issue. A positive change indicates a depreciation of the dollar.

Many international studies of equity returns include an interest rate variable as a source of risk (see for example, Ferson and Harvey (1994, 1995)). In this paper, the interest rate variable is calculated as the difference in the yield spread between a three month Eurodollar T bill and a three month U.S. T bill (the "TED spread"). The yield spread is an interesting variable to include because movements in the yield curve are often a good predictor of future economic performance. Fluctuations in this TED spread may capture fluctuations in global credit risks (Ferson and Harvey (1994, 1995)). The exchange rate series and interest rate data are available from the Federal Reserve Board of St. Louis (http://research.stlouisfed.org/fred2/). The random error term in equation (1), ε, is assumed to be normally distributed with zero mean and a generalized autoregressive conditional heteroscedasticity (GARCH) conditional variance. In equation (1) it is only unanticipated changes in interest rates, oil prices and exchange rates that should have any effect on stock returns. Since the log levels of these series are very close to a random walk the monthly rates of change of these variables are essentially unpredictable and consequently used in the estimation of equation (1).

Summary statistics for oil prices, oil price volatility, interest rates, and the trade weighted exchange rate are shown in Table 2a. The results from testing for unit roots for these variables are shown in Table 2c. Summary statistics for country specific exchange rates (country currency per $1 U.S.) are shown in Table 2b and the associated outcome from testing for unit roots shown in Table 2d. Unit root tests indicate that while each individual series is non-stationary (tables not shown but available upon request), the returns of each series are stationary.

For many financial time series, an unexpected drop in price tends to increase volatility more than an unexpected rise in price of the same magnitude (Engle and Ng, 1993). This asymmetric effect is not captured by the standard GARCH model. Nelson's (1991) exponential GARCH (EGARCH) model is designed to capture the leverage effect.[1] In the EGARCH model, the natural logarithm of the conditional variance varies across time as a function of lagged error terms. The EGARCH model for conditional variance may be written as,

$$\ln(\sigma_t^2) = \omega + \beta \ln(\sigma_{t-1}^2) + \alpha \left| \frac{\varepsilon_{t-1}}{\sigma_{t-1}} \right| + \gamma \frac{\varepsilon_{t-1}}{\sigma_{t-1}} \tag{2}$$

The order of the EGARCH model is set at (1,1) because as suggested by Bollerslev, Chou and Kroner (1992) low order GARCH models (of order (1,1)) are the most useful in practice. The EGARCH model is estimated, using a Marquardt algorithm, assuming a normal distribution for the error term. As this assumption is likely to be too strong in practice, a heteroskedasticity consistent covariance matrix is used to compute quasi maximum likelihood standard errors.

4.0 Empirical Results

Empirical results are presented using two measures of the exchange rate variable. In the first set of results (Tables 3 – 8), individual country exchange rates are used. In the second set of results, a trade weighted exchange rate variable is used (Tables 9 – 14).

4.1. Empirical Results with Country Specific Exchange Rates

Table 3 reports EGARCH(1,1) regression results that investigate the effect of oil prices on Asia Pacific equity markets. Overall, the regression model fit varies from poor (adjusted R squared value of 0.02 for Pakistan and Sri Lanka) to good (adjusted R squared value of 0.55 for Japan). The significance of each regression model is tested using an F statistic. The reported F statistics indicate that, except for the models pertaining to the equity markets in Pakistan and Sri Lanka all other regression models are statistically significant at the 5% level. All reported hypothesis testing in this paper uses a 5% level of significance. For the conditional variance equation, the estimated coefficient on the lagged volatility variable is statistically significant in all of the models except the stock return models of India, Japan, and Singapore.

1 Volatility asymmetry could also be attributed to the risk premium. The basic idea is that news of increasing volatility reduces the demand for a stock because of risk aversion. This decline in stock value is then followed by increased volatility as forecast by the news.

Table 3. The effect of oil prices on Asia Pacific equity markets

	Australia	China	Hong Kong	India	Indonesia	Japan	Korea	Malaysia	New Zealand	Pakistan	Philippines	Singapore	Sri Lanka	Taiwan	Thailand
Constant	-0.04	-0.65	1.72*	-0.08	2.00	-0.91*	-0.94	0.32	0.17	2.50*	0.75	-0.49	1.18	-0.35	0.39
	-0.14	-0.56	3.69	-0.06	1.22	-2.31	-1.84	0.69	0.38	2.29	1.02	-1.22	1.27	-0.37	0.51
Market	0.83*	1.01*	1.04*	0.66*	0.94*	1.11*	1.20*	0.70*	0.79*	0.35	0.76*	0.93*	0.48*	1.00*	1.09*
	14.50	10.85	11.44	4.58	4.12	14.91	10.94	7.09	7.81	1.73	4.84	11.71	2.62	6.89	6.19
Irate	0.05	3.07	-0.25	-7.36	-10.39	1.17	1.16	-3.34	0.05	-15.70*	2.48	-1.32	-3.70	-1.13	-9.64*
	0.03	0.51	-0.13	-1.44	-1.45	0.70	0.46	-1.24	0.02	-2.16	0.68	-0.65	-0.77	-0.30	-2.79
FX	-0.49*	-0.52*	-6.08*	-1.46*	-0.89*	-0.42*	-0.82*	-1.67*	-0.80*	-1.15*	-1.40*	-0.85*	-2.25*	-1.38*	-0.62
	-5.05	-3.85	-2.31	-2.74	-5.07	-3.91	-2.16	-5.07	-5.37	-2.49	-3.98	-2.67	-3.62	-3.86	-1.57
DUM97	0.55	1.15	-2.20*	0.82	-1.88	0.54	0.40	-0.33	-0.55	-1.73	-0.58	0.64	0.47	0.05	-0.91
	1.50	0.98	-3.17	0.60	-0.83	0.94	0.63	-0.61	-0.84	-1.61	-0.57	1.41	0.41	0.05	-0.95
Oil price	0.05*	0.16*	-0.01	0.20*	0.05	0.09*	0.07	0.05	0.04	-0.14	-0.07	0.00	-0.24*	-0.02	-0.03
	2.09	3.09	-0.29	3.12	0.46	2.94	1.53	1.07	0.96	-1.58	-1.30	-0.10	-3.83	-0.45	-0.42
ω	0.10	-0.27*	0.03	5.85*	1.10	2.14	-0.01	-0.14	0.97	0.20	0.20	2.83*	0.57*	-0.03	0.03
	1.04	-2.00	0.22	3.82	0.77	1.58	-0.12	-0.91	1.17	1.38	0.55	4.01	4.61	-0.82	0.36
α	-0.09	0.52*	0.26*	0.13	0.15	0.38*	0.00	0.45*	0.42*	-0.10	0.10	0.91*	-0.27*	-0.01	0.12
	-0.81	3.80	2.63	0.87	1.62	2.51	-0.04	3.51	3.09	-0.84	1.13	5.65	-3.20	-0.55	1.81
γ	-0.06	0.05	-0.17	-0.13	-0.17	-0.05	-0.14*	-0.04	-0.11	-0.09*	-0.02	0.16	0.06	-0.01	-0.08
	-1.49	0.59	-1.47	-1.38	-1.49	-0.47	-2.21	-0.45	-1.16	-2.99	-0.33	1.34	0.94	-0.21	-1.75
β	0.98*	0.97*	0.93*	-0.51	0.76*	0.16	1.00*	0.94*	0.60*	0.97*	0.93*	-0.13	0.92*	1.01*	0.97*
	116.84	26.09	25.02	-1.37	2.70	0.32	102.55	21.26	2.24	63.04	9.74	-0.68	44.69	171.60	57.82
Adj R squared	0.42	0.13	0.31	0.17	0.21	0.55	0.24	0.28	0.28	0.02	0.24	0.41	0.02	0.14	0.19
DW	2.39	1.78	1.78	2.10	2.00	2.07	1.98	1.79	2.33	1.98	1.78	1.85	2.01	1.81	1.94
F	19.52*	3.74*	12.29*	4.70*	7.90*	32.09*	8.90*	10.74*	11.00*	1.35	8.77*	18.75*	1.31	4.94*	7.02*
SC–LM	29.18*	8.81	13.60	19.93	17.53	13.21	13.68	12.02	11.03	15.66	15.22	17.25	17.88	21.28*	16.25
ARCH-LM	12.10	10.12	5.10	10.98	85.30*	9.31	12.46	4.85	7.44	8.74	9.66	33.50*	12.03	7.71	3.82

EGARCH(1,1) model estimation. z statistics, calculated from robust standard errors, shown below coefficient estimates. * $p < 0.05$.

Perhaps a bit surprising is the fact the estimated coefficient on the leverage term is statistically significant in only two (South Korea and Pakistan) of the estimated models indicating that stock return leverage effects are not that important in most of the countries studied. Overall, the empirical results presented in this paper, in terms of model fit and hypotheses testing are not very sensitive to whether estimation is carried out using an EGARCH model with leverage or an EGARCH model without leverage. The resulting tables in this paper, present empirical results using an EGARCH model with leverage because at least some of the equity return data contains leverage effects.

Lagrange multiplier (LM) diagnostic tests for twelve order serial correlation (SC – LM) and twelve order ARCH effects (ARCH – LM) are presented. The residuals from the models estimated for Australia and Taiwan exhibit evidence of serial correlation at the 5% level of significance. Only two models (Indonesia, Singapore) exhibits further ARCH effects. A comparison between the plot of residuals from the Indonesian stock returns model with residuals from stock returns models for other countries revealed that for most countries the residuals exhibit a lot of volatility between 1997 and 2001. In the case of Indonesia, the residuals exhibit a lot of volatility between 1997 and 2001 as well as between 1988 and 1992. Thus Indonesian stock returns have two distinct periods of high volatility. This interesting observation is not further pursued in this present paper but does present a direction for future research.

The estimated market beta is positive and statistically significant in fourteen out of the fifteen countries studied. The estimated market beta values for China, Hong Kong, Japan, South Korea, and Thailand are each statistically significant and greater than unity indicating that the equity markets in these countries are riskier than the world equity market.

The estimated oil price beta is statistically significant for five of the countries studied supporting the conjecture that oil price changes are priced risk factors. Oil price increases have a positive and significant impact on the equity markets in Australia, China, India, and Japan. This is a bit surprising since, with the exception of Australia, there countries are net importers of oil. One possible explanation for the significant and positive coefficient on the oil price variable is that rising oil prices coincide with increased economic activity (both locally and globally) and stock market strength. Oil price increases have a negative and statistically significant impact on the equity market in Sri Lanka. Oil price changes have little impact on the Indonesian equity market which is a bit surprising since Indonesia is a member of OPEC. This results is not too concerning because, as indicated in the following tables, oil price volatility does have a statistically significant impact on the Indonesian equity market.

The exchange rate variable is negative and statistically significant in fourteen of the fifteen country models studied. For each country, an increase in the amount of country currency purchased per U.S. dollar (a domestic country depreciation of its currency) has a negative and statistically significant impact on the equity market (which is measured in U.S. dollars in this study).

In general, movements in interest rates have little impact on Asia Pacific equity markets. Interest rate increases do have a negative and statistically significant impact on the equity markets in Pakistan and Thailand.

A dummy variable (DUM97), which takes on the value of one for the months after September 1997 (and zero otherwise), is included to capture structural changes that may have arisen from the Asian economic and financial crises. Surprisingly, this dummy variable is negative and statistically significant only in the case of Hong Kong stock returns. Apparently, the stock returns in the other countries did not suffer any lasting impacts from the Asian financial crisis.

Table 4 reports EGARCH(1,1) regression results that investigate the effect of oil prices and oil price volatility on Asia Pacific equity markets. The monthly oil price volatility series is constructed from the square root of the sum of squared daily returns over each month. Overall, the results in Table 4 are very similar, in terms of model fit and coefficient significance, to the results in Table 3. The estimated coefficient on the oil price volatility variable is statistically significant in only two cases (the equity markets for Indonesia and Sri Lanka). For both of these equity markets, increases in oil price volatility reduce equity returns.

The possibility that oil price changes have asymmetric effects on equity markets is investigated through the empirical results in Table 5. Asymmetrical oil price changes are constructed by separating oil price changes into a separate time series for positive changes (Oil up) and a separate time series for negative changes (Oil down). This separation allows for a more detailed analysis of how oil price changes actually affect equity markets. The estimated coefficient on the Oil up variable has a positive and statistically significant coefficient in the case of Australia, China, India, and Pakistan. The estimated coefficient on the Oil down variable has a positive and statistically significant coefficient in the case of Japan and South Korea. The estimated coefficient on the Oil down variable has a negative and statistically significant coefficient in the case of Hong Kong, Pakistan and Sri Lanka. Pakistan is the only country where asymmetric oil price changes (both up oil price changes and down oil price changes) have a statistically significant impact on equity returns. A t test to see if the oil price change effect was symmetric turned up a p value of 0.11 not rejecting the hypotheses that positive oil price changes have the same impact on equity markets as do negative oil price changes (Table 7). In the case of Pakistan, oil price changes have symmetrical effects on equity returns. The null hypothesis that oil price changes have symmetric effects on stock returns is rejected at the 5% level of significance for the models pertaining to India, Japan, South Korea, and Sri Lanka (Table 7).

The impact of asymmetric oil price volatility changes on Asia Pacific equity markets is investigated in Table 6. Asymmetric oil volatility is measured by creating one time series (Oil vol up) that matches oil price volatility with positive oil price changes and a second time series (Oil vol down) that matches oil price volatility with negative oil price changes. In this way, oil price volatility is separated into two separate time series, one that corresponds to positive oil price changes, and a second time series that corresponds to negative oil price changes. Overall, asymmetrical oil price volatility changes do not have much impact on most of the Asia Pacific equity markets. The estimated coefficient on the oil price volatility up variable is positive and statistically significant in the case of Indonesia. The estimated coefficient on the oil price volatility down variable is negative and statistically significant in the case of Indonesia, New Zealand, Singapore, and Sri Lanka. Only in the cases of Singapore and Sri Lanka is the null hypothesis of symmetric oil price volatility effects rejected at the 5% level of significance (Table 8). Changes in oil prices impact some Asia Pacific equity markets even after controlling for asymmetrical oil price volatility effects. The estimated coefficient on the oil price variable is positive and statistically significant for the case of Australia and Japan. The estimated coefficient on the oil price variable is negative and statistically significant for the case of Indonesia, Pakistan, and Sri Lanka.

Table 4. The effect of oil prices and oil price volatility on Asia Pacific equity markets

	Australia	China	Hong Kong	India	Indonesia	Japan	Korea	Malaysia	New Zealand	Pakistan	Philippines	Singapore	Sri Lanka	Taiwan	Thailand
Constant	0.13	0.65	2.44*	1.63	3.14*	-0.29	0.04	0.66	1.49	6.49	-0.05	0.39	4.34*	1.17	1.72
	0.25	0.35	2.59	0.92	2.53	-0.37	0.06	0.86	1.44	1.64	-0.04	0.57	2.25	0.78	1.27
Market	0.83*	1.07*	1.03*	0.65*	1.42*	1.13*	1.22*	0.70*	0.78*	0.43*	0.76*	0.93*	0.57*	1.00*	1.08*
	14.11	9.90	11.73	4.57	10.48	13.74	11.89	7.11	8.95	2.13	4.81	11.33	2.80	7.15	6.38
Irate	-0.59	4.36	-0.31	-7.23	-14.32*	1.23	-1.05	-3.43	0.29	-10.08	1.92	-1.04	-2.95	0.86	-9.35*
	-0.35	0.75	-0.15	-1.39	-3.93	0.74	-0.40	-1.26	0.10	-1.24	0.57	-0.48	-0.55	0.24	-2.78
FX	-0.49*	-0.51*	-5.34*	-1.46*	-0.83*	-0.42*	-0.69*	-1.68*	-0.80*	-1.22*	-1.41*	-0.97*	-1.84*	-1.38*	-0.63
	-5.00	-3.81	-2.12	-2.92	-9.78	-3.92	-2.63	-5.06	-5.11	-2.60	-3.94	-2.88	-2.93	-3.93	-1.61
DUM97	0.48	2.04	-2.02*	1.53	0.06	0.74	0.52	-0.31	-0.01	2.99	-0.53	0.69	1.55	0.28	-0.83
	1.08	1.60	-3.00	1.09	0.05	1.24	1.51	-0.57	-0.01	1.65	-0.52	1.41	1.02	0.25	-0.91
Oil price	0.05*	0.15*	-0.02	0.18*	-0.24*	0.07	0.09	0.05	0.02	-0.09	-0.06	-0.01	-0.25*	-0.05	-0.04
	2.05	2.50	-0.63	2.75	-3.87	1.94	1.90	0.94	0.43	-0.80	-1.08	-0.26	-3.30	-0.86	-0.58
Oil vol	-0.02	-0.22	-0.09	-0.23	-0.26*	-0.08	-0.09	-0.04	-0.16	-0.81	0.08	-0.11	-0.47*	-0.18	-0.13
	-0.31	-1.12	-1.30	-1.29	-2.19	-1.00	-1.13	-0.58	-1.48	-1.72	0.74	-1.76	-2.12	-1.16	-1.01
ω	0.10	-0.17	0.04	6.07*	8.24*	2.15	0.02	-0.13	0.67	6.94*	0.21	2.76*	0.59*	-0.02	0.04
	0.91	-1.20	0.31	4.70	36.21	1.76	0.24	-0.85	0.94	5.80	0.58	3.56	3.05	-0.45	0.52
α	-0.07	0.50*	0.27*	0.07	0.53*	0.43*	-0.03	0.44*	0.34*	0.30*	0.12	0.84*	-0.21*	-0.03	0.11
	-0.60	3.78	2.69	0.51	4.20	2.84	-0.33	3.42	2.24	2.45	1.21	5.41	-2.94	-0.82	1.73
γ	-0.05	0.03	-0.17	-0.14	-0.10	-0.05	-0.18*	-0.04	-0.02	-0.15	-0.02	0.15	0.08*	-0.02	-0.09
	-1.46	0.30	-1.53	-1.47	-1.21	-0.52	-4.46	-0.50	-0.20	-1.82	-0.30	1.14	2.27	-0.31	-1.83
β	0.98*	0.94*	0.93*	-0.55	-0.81*	0.14	1.00*	0.94*	0.71*	-0.53*	0.93*	-0.09	0.91*	1.01*	0.97*
	115.68	22.61	23.11	-1.73	-28.36	0.32	132.77	21.22	3.06	-2.25	9.82	-0.43	27.45	143.00	58.36
Adj R squared	0.42	0.13	0.31	0.17	0.16	0.55	0.24	0.28	0.29	0.03	0.23	0.42	0.02	0.13	0.19
DW	2.39	1.80	1.78	2.07	2.04	2.08	1.98	1.79	2.38	1.97	1.78	1.82	1.97	1.79	1.94
F	17.53*	3.47*	11.22*	4.33*	5.41*	28.68*	8.19*	9.76*	10.29*	1.55	7.86*	17.73*	1.29	4.45*	6.41*
SC–LM	28.25*	9.05	13.07	19.66	30.86*	12.47	14.63	13.05	13.88	15.83	14.83	14.91	20.23	21.26*	17.63
ARCH-LM	11.33	11.05	4.47	12.62	40.12*	9.32	15.10	5.20	8.42	18.46	10.01	31.01*	15.00	5.31	3.39

EGARCH(1,1) model estimation. z statistics, calculated from robust standard errors, shown below coefficient estimates. * $p < 0.05$.

Table 5. The effect of asymmetric oil prices on Asia Pacific equity markets

	Australia	China	Hong Kong	India	Indonesia	Japan	Korea	Malaysia	New Zealand	Pakistan	Philippines	Singapore	Sri Lanka	Taiwan	Thailand
Constant	0.05	0.21	2.16*	1.53	3.06	-0.46	0.21	0.72	1.46	2.19	0.22	1.14	4.18	1.24	1.75
	0.09	0.10	2.38	0.89	1.71	-0.60	0.29	0.94	1.46	1.17	0.17	1.94	1.80	0.84	1.29
Market	0.82*	1.15*	1.03*	0.63*	0.94*	1.14*	1.25*	0.70*	0.78*	0.10	0.84*	0.95*	0.50*	1.01*	1.08*
	14.71	9.24	11.59	4.27	3.76	14.64	12.09	7.15	8.85	0.57	5.29	11.15	2.05	7.30	6.36
Irate	-0.86	1.27	0.01	-7.30	-9.91	1.57	-0.49	-3.13	0.29	-12.68	1.64	0.94	-2.83	1.65	-9.24*
	-0.50	0.21	0.00	-1.38	-1.50	0.95	-0.18	-1.18	0.10	-1.85	0.39	0.47	-0.52	0.46	-2.74
FX	-0.49*	-0.53*	-6.12*	-1.40*	-0.83*	-0.44*	-0.69*	-1.68*	-0.79*	-0.96*	-1.54*	-0.81*	-1.89*	-1.32*	-0.64
	-5.03	-4.61	-2.30	-2.85	-7.10	-4.10	-2.76	-5.05	-5.10	-2.12	-4.68	-3.09	-2.97	-3.86	-1.63
DUM97	0.42	1.62	-1.83*	1.47	-0.57	0.88	0.64	-0.29	-0.03	-1.23	-0.71	0.27	0.29	0.29	-0.77
	0.97	1.29	-2.57	1.06	-0.37	1.50	1.72	-0.52	-0.04	-1.19	-0.66	0.51	0.21	0.26	-0.81
Oil up	0.12*	0.23*	0.11	0.27*	0.13	-0.02	-0.03	0.00	0.03	0.37*	0.03	0.05	0.00	-0.08	-0.07
	2.95	2.14	1.77	2.41	0.79	-0.24	-0.27	-0.05	0.38	2.56	0.32	0.74	0.01	-0.72	-0.55
Oil down	-0.04	0.00	-0.20*	0.09	-0.30	0.20*	0.24*	0.11	0.00	-0.64*	-0.11	-0.05	-0.49*	-0.01	0.01
	-0.76	0.01	-2.13	0.59	-1.36	3.06	2.23	1.13	0.03	-3.64	-0.90	-0.85	-2.77	-0.06	0.05
Oil vol	-0.06	-0.21	-0.19*	-0.28	-0.36	0.01	-0.01	0.00	-0.16	-0.39	0.00	-0.18*	-0.54	-0.16	-0.11
	-0.98	-1.03	-2.02	-1.44	-1.72	0.15	-0.10	-0.06	-1.35	-1.74	0.03	-2.45	-1.94	-0.98	-0.66
ω	0.12	-0.21*	0.04	5.98*	7.54*	2.36	0.02	-0.14	0.67	0.45*	3.52	2.42*	0.57	-0.02	0.03
	1.25	-2.17	0.30	4.62	5.44	1.59	0.24	-0.97	0.95	3.18	0.67	3.27	1.85	-0.35	0.39
α	-0.09	0.33*	0.29*	0.03	0.37*	0.33*	-0.04	0.44*	0.34*	-0.20	-0.03	0.87*	-0.07	-0.04	0.12
	-0.83	2.42	3.02	0.19	2.29	2.12	-0.46	3.39	2.25	-1.66	-0.16	5.52	-0.80	-0.92	1.72
γ	-0.05	0.06	-0.14	-0.16	-0.24*	-0.04	-0.17*	-0.04	-0.02	-0.09*	0.10	0.15	0.10*	-0.02	-0.08
	-1.42	0.70	-1.24	-1.61	-2.69	-0.39	-4.44	-0.46	-0.25	-2.83	0.82	1.08	2.33	-0.31	-1.77
β	0.98*	0.99*	0.92*	-0.52	-0.58*	0.08	1.00*	0.94*	0.71*	0.94*	0.16	0.01	0.88*	1.01*	0.97*
	139.06	37.28	20.47	-1.60	-2.51	0.16	132.99	21.67	3.06	59.12	0.13	0.03	14.43	133.43	59.82
Adj R squared	0.42	0.14	0.34	0.16	0.21	0.56	0.23	0.27	0.29	0.10	0.23	0.42	0.02	0.12	0.19
DW	2.39	1.80	1.85	2.05	2.04	2.06	1.93	1.77	2.38	1.99	1.81	1.82	1.97	1.79	1.93
F	16.01*	3.54*	11.62*	3.95*	6.46*	27.24*	7.28*	8.69*	9.29*	2.62*	7.21*	16.03*	1.31	3.94*	5.74*
SC–LM	28.05*	8.35	10.99	19.56	17.59	13.12	15.04	13.35	14.09	17.24	13.57	18.58	20.96	22.41*	17.92
ARCH-LM	11.38	9.12	4.34	11.68	74.58*	8.84	16.20	3.91	8.56	9.21	10.32	29.59*	14.94	5.14	3.56

EGARCH(1,1) model estimation. z statistics, calculated from robust standard errors, shown below coefficient estimates.* $p < 0.05$

Table 6. The effect of asymmetric oil price volatility on Asia Pacific equity markets

	Australia	China	Hong Kong	India	Indonesia	Japan	Korea	Malaysia	New Zealand	Pakistan	Philippines	Singapore	Sri Lanka	Taiwan	Thailand
Constant	0.25	0.11	2.25*	1.56	-0.57	-0.16	-0.09	0.61	1.15	2.32	-0.32	0.65	4.85*	0.84	1.54
	0.50	0.06	2.38	0.88	-0.37	-0.21	-0.12	0.80	1.50	1.03	-0.25	0.85	3.23	0.51	1.10
Market	0.83*	1.09*	1.05*	0.65*	1.05*	1.13*	1.22*	0.72*	0.81*	0.32	0.76*	0.94*	0.50*	0.99*	1.08*
	13.88	9.58	11.40	4.59	4.01	13.45	12.08	7.44	9.51	1.66	4.88	9.03	2.78	6.95	6.37
Irate	-0.25	4.25	0.39	-7.00	-9.67	1.07	-0.51	-2.25	1.00	-14.42*	2.33	-0.84	0.43	0.55	-9.15*
	-0.15	0.72	0.21	-1.34	-1.59	0.63	-0.20	-0.78	0.35	-2.04	0.72	-0.45	0.09	0.15	-2.74
FX	-0.49*	-0.52*	-6.91*	-1.46*	-1.00*	-0.40*	-0.69*	-1.65*	-0.77*	-0.99*	-1.41*	-1.13*	-1.71*	-1.34*	-0.64
	-4.98	-4.36	-2.65	-2.89	-5.53	-3.72	-2.60	-5.19	-6.25	-2.20	-3.79	-2.49	-2.65	-3.90	-1.61
DUM97	0.60	1.84	-2.26*	1.54	-1.64	0.72	0.44	-0.23	-0.42	-0.32	-0.47	0.96	1.25	0.37	-0.82
	1.32	1.42	-3.60	1.09	-1.09	1.24	1.17	-0.41	-0.60	-0.31	-0.46	1.73	1.00	0.32	-0.88
Oil price	0.09*	0.16	-0.11	0.15	-0.58*	0.13*	0.03	-0.08	-0.05	-0.34*	-0.20	-0.08	-0.56*	-0.16	-0.09
	2.22	1.71	-1.70	1.40	-2.99	2.10	0.30	-1.08	-0.93	-2.18	-1.94	-1.17	-4.85	-1.51	-0.69
Oil vol up	-0.08	-0.15	0.05	-0.19	0.87*	-0.15	0.00	0.08	-0.01	0.05	0.26	-0.06	-0.19	-0.04	-0.06
	-1.22	-0.65	0.52	-0.94	2.36	-1.85	0.02	0.75	-0.14	0.21	1.53	-0.62	-0.94	-0.16	-0.32
Oil vol down	0.00	-0.13	-0.13	-0.26	-0.31*	-0.03	-0.13	-0.13	-0.20*	-0.33	-0.03	-0.27*	-0.81*	-0.27	-0.17
	0.05	-0.70	-1.67	-1.34	-1.97	-0.36	-1.36	-1.75	-2.44	-1.16	-0.25	-3.60	-4.24	-1.69	-1.12
ω	0.09	-0.19	0.01	6.05*	0.13	2.06	0.02	-0.13	3.57*	0.29*	0.17	1.41*	0.66*	-0.04	0.03
	0.82	-1.59	0.05	4.86	0.63	1.46	0.21	-0.87	5.11	2.26	0.60	2.22	2.60	-0.90	0.38
α	-0.07	0.45*	0.26*	0.08	0.27*	0.38*	-0.03	0.43*	0.50*	-0.13	0.14	0.67*	-0.25*	-0.01	0.11
	-0.57	3.23	2.74	0.63	2.59	2.43	-0.36	3.37	3.72	-1.08	1.48	4.36	-2.98	-0.28	1.74
γ	-0.05	0.04	-0.16	-0.14	-0.18*	-0.05	-0.18*	-0.04	-0.24*	-0.11*	-0.02	0.23*	0.14*	-0.03	-0.08
	-1.26	0.43	-1.53	-1.52	-2.02	-0.51	-4.41	-0.50	-2.50	-3.12	-0.23	2.41	2.78	-0.48	-1.77
β	0.98*	0.96*	0.94*	-0.55	0.93*	0.18	1.00*	0.94*	-0.22	0.96*	0.93*	0.37	0.90*	1.01*	0.97*
	116.13	25.95	28.36	-1.80	22.62	0.35	132.52	19.97	-1.11	61.59	12.24	1.85	20.19	167.34	60.87
Adj R squared	0.42	0.12	0.31	0.16	0.19	0.55	0.24	0.30	0.30	0.03	0.23	0.43	0.05	0.13	0.19
DW	2.38	1.79	1.77	2.08	2.05	2.09	2.00	1.79	2.36	1.99	1.80	1.79	1.94	1.80	1.95
F	15.81*	3.11*	10.15*	3.89*	5.96*	26.57*	7.40*	9.78*	9.79*	1.45	7.23*	16.40*	1.75	4.08*	5.85*
SC–LM	27.99*	8.70	14.54	20.25	22.76*	14.56	15.41	13.38	19.25	14.21	15.74	10.58	20.18	23.16*	17.98
ARCH-LM	11.03	10.65	4.71	12.35	69.72*	9.42	16.20	4.75	9.46	7.83	8.02	27.22*	14.37	6.41	3.47

EGARCH(1,1) model estimation. z statistics, calculated from robust standard errors, shown below coefficient estimates. * $p < 0.05$

4.2. Empirical Results with Trade Weighted Exchange Rates

In this section of the paper, the sensitivity of the results presented in the previous section of the paper is compared to a different measure of exchange rates. In particular, country specific exchange rates are replaced with a trade weighted exchange rate variable.

While the numerical values from the regression output reported in Table 9 vary from the numerical values of the regression output reported in Table 3, there are some qualitative similarities. Overall, the regression model fit in Table 9 varies from poor (adjusted R squared value of -0.01 for Pakistan) to good (adjusted R squared value of 0.52 for Japan) and this range of adjusted R squared values is very similar to the range of adjusted R squared values reported in Table 3. The estimated market beta values shown in Table 9 are positive and statistically significant in thirteen out of the fifteen countries studied (compared to fourteen out the fifteen countries studied in Table 3). The estimated coefficient on the oil price variable is statistically significant in six of the countries studied (which includes the same five countries from Table 3 where the estimated coefficient on the oil price variable is statistically significant). The biggest difference between the results reported in the two tables is that in Table 3 the estimated coefficient on the foreign exchange rate variable is statistically significant in the model for every country studied, while in Table 9, the estimated coefficient on the trade weighted exchange rate variable is statistically significant in only two of the models. As in the case of Table 3, the interest rate variable has little power in explaining stock price movements.

Comparing the regression results reported in Tables 10 – 12 to their respective corresponding Tables 4 – 6 reveals qualitatively similar results. There is little evidence suggesting that oil price volatility (either symmetric or asymmetric) helps to explain stock returns in Asia Pacific equity markets. When the trade weighted exchange rate variable is used, there is some evidence supporting the conjecture that oil prices have asymmetric effects on stock returns (comparing Table 11 to Table 5). In the cases of Australia, India, Japan, Korea, the Philippines, and Sri Lanka, evidence is found showing that oil prices have an asymmetric effect on stock prices (Table 13). Only in the case of Singapore, is evidence found that oil price volatility has an asymmetric effect on stock prices (Table 14). Comparing the results in Table 7 with the results in Table 13, oil prices have asymmetric effects on the stock returns in India, Japan, Korea, and Sri Lanka. Comparing the results in Table 8 with the results in Table 14, oil price volatility has asymmetric effects on the stock returns in Singapore.

Table 7. Wald tests of oil price symmetry

	Australia	China	Hong Kong	India	Indonesia	Japan	Korea	Malaysia	New Zealand	Pakistan	Philippines	Singapore	Sri Lanka	Taiwan	Thailand
F	3.27	3.47	1.36	7.65	0.86	7.95	5.12	1.11	0.17	2.61	0.42	0.01	9.42	0.65	0.25
p value	0.07	0.06	0.25	0.01	0.36	0.01	0.02	0.29	0.68	0.11	0.52	0.94	0.00	0.42	0.62

Table 8. Wald tests of oil price volatility symmetry

	Australia	China	Hong Kong	India	Indonesia	Japan	Korea	Malaysia	New Zealand	Pakistan	Philippines	Singapore	Sri Lanka	Taiwan	Thailand
F	0.48	0.53	0.30	1.61	1.96	1.62	0.56	0.19	2.34	0.34	0.94	5.56	9.99	0.83	0.68
p value	0.49	0.47	0.58	0.21	0.16	0.20	0.46	0.67	0.13	0.56	0.33	0.02	0.00	0.36	0.41

Table 9. The effect of oil prices on Asia Pacific equity markets (TWE)

	Australia	China	Hong Kong	India	Indonesia	Japan	Korea	Malaysia	New Zealand	Pakistan	Philippines	Singapore	Sri Lanka	Taiwan	Thailand
Constant	-0.09	-0.23	1.68*	-0.34	3.26	-0.76	-0.69	0.47	0.63	0.55	2.26*	0.30	0.19	-1.36	0.57
	-0.27	-0.20	3.88	-0.32	1.64	-1.68	-1.31	0.89	1.42	0.54	2.99	0.82	0.18	-1.30	0.70
Market	0.90*	0.95*	1.01*	0.67*	1.04*	1.14*	1.23*	0.68*	0.80*	0.37	0.71*	1.00*	0.26	1.11*	1.11*
	16.11	10.46	11.09	4.78	3.40	15.30	10.94	6.43	9.55	1.75	4.82	11.24	1.09	7.35	6.46
Irate	-1.41	2.84	1.01	-11.08*	-6.36	1.87	-1.04	-5.02	-0.79	-15.99*	-1.74	1.48	1.40	0.00	-9.28*
	-0.75	0.45	0.53	-2.14	-0.87	1.05	-0.41	-1.92	-0.27	-2.20	-0.54	0.78	0.24	0.00	-2.66
TWE	0.03	-1.00	-0.22	-0.21	-2.21*	-0.37	-0.27	-0.32	-0.85*	0.84	-0.54	-0.03	-0.83	0.45	0.41
	0.14	-1.84	-0.70	-0.38	-2.22	-1.38	-0.64	-1.05	-3.46	1.32	-1.10	-0.13	-1.25	1.09	0.75
DUM97	0.61	0.77	-2.22*	0.91	-3.05	0.38	0.42	-0.53	-1.01	-0.05	-3.55*	-0.05	0.65	0.91	-1.25
	1.62	0.63	-3.48	0.74	-1.21	0.60	0.99	-0.83	-1.52	-0.04	-3.80	-0.10	0.43	0.75	-1.30
Oil price	0.07*	0.13*	-0.01	0.20*	0.08	0.10*	0.12*	0.07	0.03	-0.11	-0.11	0.04	-0.20*	0.00	-0.05
	2.80	2.45	-0.43	2.84	0.54	3.52	2.37	1.42	0.72	-1.29	-1.83	1.25	-2.27	-0.09	-0.71
ω	0.12	-0.27*	0.02	5.70*	3.02*	1.71	0.05	-0.11	3.30*	0.27	0.25	2.19*	2.92	-0.05	0.09
	1.01	-2.56	0.18	3.65	2.58	1.80	0.78	-0.63	5.94	1.86	1.57	3.64	0.75	-1.19	1.26

Table 9. Continued

	Australia	China	Hong Kong	India	Indonesia	Japan	Korea	Malaysia	New Zealand	Pakistan	Philippines	Singapore	Sri Lanka	Taiwan	Thailand
α	-0.08	0.43*	0.27*	-0.19	0.68*	0.45*	-0.05	0.49*	0.66*	-0.12	-0.12	1.01*	0.23	-0.01	0.06
	-0.74	4.06	2.60	-0.85	4.48	2.75	-0.67	3.02	4.52	-1.01	-1.45	6.47	1.46	-0.41	1.05
γ	-0.07*	0.08	-0.18	-0.18	-0.41*	-0.10	-0.17*	-0.05	-0.32*	-0.10*	-0.08	0.14	-0.07	-0.04	-0.10*
	-3.04	1.01	-1.60	-1.45	-3.35	-0.97	-4.87	-0.52	-3.20	-3.43	-1.48	0.95	-0.47	-0.90	-2.20
β	0.98*	0.98*	0.93*	-0.39	0.30	0.30	1.00*	0.93*	-0.14	0.96*	0.96*	0.07	0.32	1.01*	0.97*
	98.35	39.18	24.61	-0.91	1.32	0.88	131.18	18.31	-0.88	62.08	23.11	0.39	0.37	134.23	64.28
Adj R squared	0.35	0.14	0.30	0.12	0.04	0.52	0.19	0.13	0.22	-0.01	0.11	0.34	0.00	0.07	0.16
DW	2.24	1.85	1.77	1.92	1.77	1.98	1.97	1.66	2.18	1.94	1.55	1.83	1.97	1.80	1.84
F	14.76*	3.95*	11.96*	3.46*	2.19*	28.81*	7.09*	4.91*	8.07*	0.80	4.09*	14.20*	0.98	2.97*	5.80*
SC–LM	16.27	7.46	14.41	22.39*	19.46	9.87	16.84	15.52	18.28	13.22	23.85*	19.72	19.63	13.04	19.57
ARCH-LM	10.80	7.79	4.48	15.99	59.58*	12.93	13.47	12.17	16.85	7.23	9.30	31.51*	7.86	8.20	5.02

EGARCH(1,1) model estimation. z statistics, calculated from robust standard errors, shown below coefficient estimates. * $p < 0.05$.

Table 10. The effect of oil prices and oil price volatility on Asia Pacific equity markets (TWE)

	Australia	China	Hong Kong	India	Indonesia	Japan	Korea	Malaysia	New Zealand	Pakistan	Philippines	Singapore	Sri Lanka	Taiwan	Thailand
Constant	-0.20	0.38	2.66*	1.25	5.05*	-0.48	0.24	0.77	1.14	4.99	2.47	1.49*	3.92	0.40	1.69
	-0.35	0.20	2.84	0.70	2.35	-0.60	0.30	0.89	1.24	1.16	1.87	2.57	1.51	0.26	1.23
Market	0.93*	0.98*	0.99*	0.69*	0.98*	1.14*	1.23*	0.68*	0.77*	0.49*	0.66*	0.99*	0.30	1.10*	1.10*
	17.70	10.18	11.17	5.05	3.53	15.38	10.72	6.45	8.70	2.53	4.32	11.48	1.28	7.47	6.67
Irate	-0.68	3.37	0.92	-12.63*	-4.35	1.88	-0.43	-5.13	-0.42	-6.16	0.28	1.52	-1.82	1.70	-9.06*
	-0.37	0.53	0.47	-2.48	-0.66	1.06	-0.16	-1.93	-0.14	-0.88	0.09	0.81	-0.32	0.48	-2.68
TWE	0.02	-0.91	-0.27	-0.13	-2.77*	-0.37	-0.23	-0.35	-0.91*	0.35	-0.47	-0.01	-1.59*	0.42	0.38
	0.13	-1.82	-0.87	-0.25	-2.80	-1.38	-0.54	-1.14	-3.50	0.45	-0.91	-0.05	-2.24	1.04	0.70
DUM97	0.67	1.50	-2.29*	1.67	-3.70	0.45	0.49	-0.47	-0.89	4.77*	-3.61*	0.34	0.46	0.97	-1.13
	1.53	1.18	-3.76	1.41	-1.45	0.71	1.13	-0.73	-1.24	2.69	-3.24	0.65	0.27	0.79	-1.23

Table 10. Continued

	Australia	China	Hong Kong	India	Indonesia	Japan	Korea	Malaysia	New Zealand	Pakistan	Philippines	Singapore	Sri Lanka	Taiwan	Thailand
Oil price	0.07*	0.12*	-0.03	0.20*	-0.02	0.10*	0.11*	0.07	0.03	-0.12	-0.13*	0.03	-0.18*	-0.02	-0.05
	2.67	2.09	-0.69	2.99	-0.15	3.30	2.26	1.20	0.85	-1.09	-2.08	1.03	-2.43	-0.42	-0.79
Oil vol	0.01	-0.12	-0.09	-0.22	-0.15	-0.03	-0.10	-0.03	-0.05	-0.84	-0.05	-0.17*	-0.39	-0.19	-0.12
	0.09	-0.67	-1.26	-1.21	-0.80	-0.51	-1.20	-0.49	-0.52	-1.68	-0.38	-2.98	-1.38	-1.23	-0.98
ω	0.15	-0.26*	0.02	5.73*	2.65*	1.76	0.04	-0.10	3.58*	6.68*	0.23	1.88*	6.54*	-0.04	0.09
	1.37	-2.65	0.18	4.37	2.40	1.84	0.55	-0.52	6.05	7.27	1.91	3.54	2.86	-0.93	1.22
α	-0.13	0.39*	0.25*	-0.27	0.63*	0.46*	-0.04	0.47*	0.57*	0.45*	-0.11	1.05*	-0.22	-0.01	0.06
	-1.09	3.62	2.73	-1.17	3.62	2.79	-0.49	2.87	3.94	3.81	-1.59	6.73	-1.04	-0.36	1.10
γ	-0.07*	0.08	-0.18	-0.19	-0.38*	-0.10	-0.16*	-0.06	-0.29*	-0.14*	-0.08	0.13	-0.01	-0.05	-0.09*
	-2.49	1.08	-1.76	-1.54	-2.79	-1.00	-4.71	-0.59	-2.84	-2.03	-1.83	0.89	-0.07	-0.89	-2.19
β	0.98*	0.99*	0.94*	-0.38	0.39	0.28	1.00*	0.92*	-0.21	-0.49*	0.96*	0.15	-0.42	1.01*	0.97*
	109.79	45.73	25.91	-1.05	1.70	0.82	143.98	18.13	-1.25	-2.85	38.61	0.92	-0.84	126.09	66.19
Adj R squared	0.35	0.13	0.30	0.11	0.05	0.52	0.19	0.13	0.22	-0.01	0.10	0.35	0.00	0.07	0.16
DW	2.25	1.86	1.76	1.90	1.76	1.99	1.96	1.66	2.20	1.90	1.56	1.79	1.93	1.77	1.84
F	13.02*	3.56*	10.93*	3.10*	2.25*	25.84*	6.28*	4.53*	7.51*	0.80	3.54*	13.18*	1.05	2.66*	5.33*
SC–LM	17.34	8.19	14.04	22.21*	20.57	9.95	16.92	16.03	20.15	17.76	26.05*	19.26	19.99	13.61	20.86
ARCH-LM	12.45	8.40	4.09	16.52	59.37*	12.72	15.67	12.43	15.51	21.39*	9.36	28.99*	10.99	6.08	5.00

EGARCH(1,1) model estimation. z statistics, calculated from robust standard errors, shown below coefficient estimates. * $p < 0.05$.

Table 11. The effect of asymmetric oil prices on Asia Pacific equity markets (TWE)

	Australia	China	Hong Kong	India	Indonesia	Japan	Korea	Malaysia	New Zealand	Pakistan	Philippines	Singapore	Sri Lanka	Taiwan	Thailand
Constant	-0.35	0.08	2.40*	1.06	2.92	-0.46	0.13	0.89	1.15	2.69	2.35	1.52*	4.18	0.38	1.97
	-0.64	0.04	2.66	0.61	1.41	-0.55	0.16	1.04	1.51	1.47	1.87	2.58	1.68	0.24	1.47
Market	0.89*	1.03*	0.99*	0.65*	0.93*	1.15*	1.26*	0.69*	0.77*	0.16	0.68*	0.98*	0.29	1.10*	1.07*
	16.19	9.64	10.92	4.60	3.39	15.55	11.06	6.84	8.58	0.88	4.93	11.58	1.19	7.45	6.83

Table 11. Continued

	Australia	China	Hong Kong	India	Indonesia	Japan	Korea	Malaysia	New Zealand	Pakistan	Philippines	Singapore	Sri Lanka	Taiwan	Thailand
Irate	-1.69	2.66	1.01	-12.92*	-7.47	2.23	0.15	-4.64	-0.32	-12.04	0.18	1.41	-1.90	1.47	-8.74*
	-0.90	0.44	0.54	-2.46	-1.13	1.22	0.05	-1.79	-0.11	-1.64	0.06	0.77	-0.35	0.41	-2.57
TWE	0.04	-0.94	-0.29	-0.04	-2.27*	-0.41	-0.23	-0.31	-0.89*	0.42	-0.51	-0.03	-1.53*	0.44	0.52
	0.21	-1.88	-0.92	-0.08	-2.35	-1.50	-0.54	-0.98	-3.36	0.74	-1.10	-0.13	-2.25	1.07	0.94
DUM97	0.60	1.71	-2.24*	1.60	-3.59	0.61	0.60	-0.40	-0.85	-0.05	-3.62*	0.31	0.20	1.07	-1.40
	1.41	1.36	-3.53	1.40	-1.37	0.97	1.40	-0.63	-1.19	-0.04	-4.16	0.59	0.12	0.85	-1.47
Oil up	0.16*	0.24*	0.09	0.37*	0.35	0.02	0.00	0.00	0.04	0.25	-0.13	0.04	0.05	-0.04	-0.13
	3.73	2.28	1.34	3.23	1.44	0.29	-0.03	-0.03	0.52	1.60	-1.20	0.65	0.40	-0.40	-0.96
Oil down	-0.04	-0.04	-0.18*	0.04	-0.30	0.20*	0.25*	0.15	0.01	-0.50*	-0.11	0.01	-0.41*	0.00	0.01
	-0.66	-0.32	-1.98	0.26	-1.30	2.86	2.24	1.44	0.16	-2.68	-1.02	0.21	-2.35	0.02	0.06
Oil vol	-0.06	-0.21	-0.17	-0.30	-0.19	0.02	0.00	0.01	-0.06	-0.45*	-0.01	-0.18*	-0.58	-0.18	-0.08
	-0.86	-1.05	-1.75	-1.60	-0.87	0.27	0.03	0.06	-0.64	-1.99	-0.06	-2.70	-1.90	-1.06	-0.53
ω	0.12	-0.24*	0.04	5.76*	2.67*	1.92	0.05	-0.12	3.73*	0.38*	0.32*	1.88*	6.80*	-0.05	0.13
	1.31	-2.38	0.29	5.53	3.41	1.71	0.72	-0.73	6.18	2.41	2.49	3.52	3.80	-1.10	1.87
α	-0.08	0.32*	0.27*	-0.35	0.71*	0.38*	-0.04	0.47*	0.54*	-0.19	-0.16*	1.03*	-0.25	-0.01	0.02
	-0.82	3.21	2.90	-1.58	4.91	2.25	-0.50	2.91	3.82	-1.60	-2.26	6.61	-1.28	-0.22	0.37
γ	-0.06*	0.09	-0.16	-0.22	-0.45*	-0.10	-0.16*	-0.05	-0.28*	-0.11*	-0.09	0.13	0.02	-0.05	-0.10*
	-2.71	1.22	-1.60	-1.69	-3.70	-0.93	-4.92	-0.53	-2.76	-3.74	-1.62	0.87	0.20	-0.96	-2.57
β	0.98*	1.00*	0.93*	-0.38	0.37*	0.24	0.99*	0.93*	-0.25	0.95*	0.95*	0.15	-0.48	1.01*	0.97*
	120.78	52.59	22.18	-1.31	2.36	0.60	143.88	19.81	-1.47	51.87	31.02	0.95	-1.25	127.67	73.71
Adj R squared	0.35	0.15	0.33	0.10	0.04	0.53	0.18	0.12	0.22	0.06	0.10	0.35	0.01	0.06	0.14
DW	2.25	1.87	1.81	1.89	1.83	1.96	1.93	1.65	2.20	1.93	1.55	1.79	1.93	1.77	1.83
F	12.06*	3.76*	11.05*	2.77*	1.95	24.27*	5.68*	3.83*	6.81*	1.95	3.27*	12.08*	1.11	2.43*	4.38*
SC–LM	16.24	8.29	12.53	22.85*	21.97*	10.67	16.50	16.82	20.55	16.49	24.55*	21.72*	21.03*	14.11	22.19*
ARCH-LM	11.43	6.58	4.06	13.83	55.07*	11.97	17.16	10.68	15.51	9.90	9.93	28.17*	10.27	6.19	4.93

EGARCH(1,1) model estimation. z statistics, calculated from robust standard errors, shown below coefficient estimates.* $p < 0.05$.

Table 12. The effect of asymmetric oil price volatility on Asia Pacific equity markets (TWE)

	Australia	China	Hong Kong	India	Indonesia	Japan	Korea	Malaysia	New Zealand	Pakistan	Philippines	Singapore	Sri Lanka	Taiwan	Thailand
Constant	-0.07	0.09	2.56*	1.23	0.48	-0.68	0.30	0.73	1.24	2.49	2.21	1.41*	2.91	-0.60	1.57
	-0.12	0.05	2.75	0.71	0.32	-0.95	0.37	0.85	1.50	0.81	1.72	2.53	1.29	-0.38	1.11
Market	0.90*	0.92*	1.00*	0.69*	0.90*	1.18*	1.23*	0.71*	0.77*	0.38	0.72*	1.00*	0.28	1.12*	1.09*
	16.26	9.84	10.97	5.05	3.61	17.57	10.54	6.94	8.43	1.76	5.10	11.82	1.22	7.81	6.75
Irate	-1.92	3.28	0.79	-12.59*	-8.16	0.41	-0.03	-1.72	-0.11	-14.29	-0.81	1.87	0.34	-0.37	-9.15*
	-1.04	0.53	0.42	-2.46	-1.39	0.28	-0.01	-0.55	-0.04	-1.69	-0.26	1.00	0.06	-0.10	-2.70
TWE	0.04	-1.09*	-0.26	-0.13	-1.39	-0.38	-0.35	-0.40	-0.88*	0.64	-0.48	0.11	-1.30*	0.38	0.35
	0.22	-2.04	-0.81	-0.25	-1.71	-1.65	-0.81	-1.25	-3.24	0.81	-1.05	0.50	-2.05	0.93	0.64
DUM97	0.56	0.83	-2.38*	1.67	-1.76	0.05	0.28	-0.23	-0.74	2.61	-3.53*	0.42	0.95	1.59	-1.15
	1.22	0.66	-4.02	1.38	-0.97	0.11	0.60	-0.34	-1.03	1.65	-4.24	0.82	0.57	1.29	-1.23
Oil price	0.13*	0.07	-0.08	0.19	-0.65*	0.19*	0.08	-0.08	-0.03	-0.18	-0.21*	-0.03	-0.53*	-0.16	-0.11
	3.14	0.77	-1.17	1.70	-3.02	3.56	0.86	-1.06	-0.52	-0.81	-2.33	-0.58	-4.32	-1.54	-0.84
Oil vol up	-0.06	0.02	-0.02	-0.21	0.80*	-0.08	-0.04	0.09	0.01	-0.31	0.11	-0.09	0.02	-0.02	-0.05
	-0.87	0.11	-0.18	-1.21	2.16	-1.13	-0.32	0.78	0.13	-1.07	0.67	-1.33	0.09	-0.08	-0.28
Oil vol down	0.05	-0.10	-0.12	-0.22	-0.47*	0.07	-0.12	-0.15	-0.15	-0.58	-0.08	-0.23*	-0.66*	-0.30	-0.15
	0.66	-0.56	-1.40	-0.99	-2.99	0.85	-1.24	-1.83	-1.66	-1.24	-0.65	-3.29	-2.22	-1.91	-1.11
ω	0.13	-0.27*	0.02	5.73*	0.32	0.00	0.02	-0.11	3.72*	5.97*	0.28*	1.80*	7.67*	-0.05	0.10
	1.09	-2.28	0.13	4.36	1.17	-0.01	0.28	-0.59	6.25	3.33	1.96	3.39	5.52	-1.10	1.28
α	-0.09	0.46*	0.24*	-0.27	0.33*	0.18	-0.04	0.45*	0.51*	0.19	-0.14	1.04*	-0.17	-0.01	0.06
	-0.74	4.14	2.67	-1.17	2.84	1.93	-0.50	2.77	3.56	1.03	-1.96	6.66	-0.87	-0.31	1.06
γ	-0.07*	0.07	-0.18	-0.19	-0.23*	-0.10	-0.16*	-0.05	-0.28*	-0.10	-0.09	0.13	0.03	-0.05	-0.10*
	-2.94	0.87	-1.82	-1.54	-2.06	-1.44	-4.29	-0.54	-2.89	-1.10	-1.40	0.90	0.37	-1.14	-2.20
β	0.97*	0.98*	0.94*	-0.38	0.89*	0.95*	1.00*	0.93*	-0.24	-0.30	0.96*	0.17	-0.69*	1.01*	0.97*
	101.81	32.54	27.70	-1.05	15.29	17.12	137.82	17.54	-1.42	-0.78	26.39	1.04	-2.21	113.30	65.91

Table 12. Continued

	Australia	China	Hong Kong	India	Indonesia	Japan	Korea	Malaysia	New Zealand	Pakistan	Philippines	Singapore	Sri Lanka	Taiwan	Thailand
Adj R squared	0.34	0.13	0.30	0.11	0.06	0.53	0.19	0.16	0.24	0.01	0.11	0.35	0.04	0.06	0.16
DW	2.22	1.87	1.75	1.90	1.82	1.98	1.97	1.66	2.21	1.95	1.57	1.80	1.93	1.77	1.85
F	11.81*	3.25*	9.85*	2.80*	2.37*	23.86*	5.72*	5.00*	7.36*	1.22	3.48*	12.05*	1.67	2.35*	4.91*
SC–LM	16.01	8.37	15.32	22.67*	21.77*	15.73	17.89	15.88	19.92	11.64	22.86*	19.40	16.40	13.51	21.00
ARCH-LM	11.27	8.66	4.32	16.46	51.14*	8.16	14.73	9.05	12.58	18.01	9.49	29.86*	10.20	5.79	4.89

EGARCH(1,1) model estimation. z statistics, calculated from robust standard errors, shown below coefficient estimates. * $p < 0.05$

Table 13. Wald tests of oil price symmetry (TWE)

	Australia	China	Hong Kong	India	Indonesia	Japan	Korea	Malaysia	New Zealand	Pakistan	Philippines	Singapore	Sri Lanka	Taiwan	Thailand
F	5.99	2.69	1.50	9.01	0.03	11.66	6.05	1.86	0.64	1.78	4.10	0.90	5.73	0.13	0.79
p value	0.02	0.10	0.22	0.00	0.86	0.00	0.01	0.17	0.43	0.18	0.04	0.34	0.02	0.72	0.38

Table 14. Wald tests of oil price volatility symmetry (TWE)

	Australia	China	Hong Kong	India	Indonesia	Japan	Korea	Malaysia	New Zealand	Pakistan	Philippines	Singapore	Sri Lanka	Taiwan	Thailand
F	0.01	0.05	0.86	1.48	0.78	0.01	0.93	0.14	0.73	1.59	0.02	7.91	1.64	0.99	0.65
p value	0.91	0.83	0.36	0.23	0.38	0.91	0.34	0.71	0.39	0.21	0.90	0.01	0.20	0.32	0.42

5. Discussion and Conclusions

This paper uses an international multifactor model to investigate the relationship between oil prices and stock prices in the Asia Pacific region of the world. This is a particularly interesting region of the world to study because the Asia Pacific region is home to some of the fastest growing economies in the world. Moreover, the Asia Pacific region includes the two most populous economies in the world (China and India) and the region contains a mix of developed economies and emerging economies. The econometric results show that for approximately half of the countries studied, oil price changes do have a significant impact on the equity returns of countries in the Asia Pacific region. Moreover, oil price changes have asymmetric impacts on equity returns in India, Japan, South Korea, and Sri Lanka and these results are robust to two different measures of the exchange rage variable. These results are important for three reasons. First, oil price changes are a priced risk factor for about half of the countries studied and a better understanding of equity price movements in the Asia Pacific region of the world can be obtained by adding an oil price variable to equity pricing models. Second, for countries in which oil price changes were found to have asymmetric effects on stock returns, it is important for investors, policy makers, and portfolio managers to realize that studying oil price shocks in one direction, say positive shocks, does not guarantee a complete understanding of the response of stock returns to oil price shocks in the opposite direction. Third, these results are important to international investors searching for ways to diversify portfolio risks. If oil prices are priced risk factors in equity pricing models, then an investment portfolio invested in equities and oil would not diversify risks in the same way as if oil and equities were uncorrelated.

This paper also investigates the relationship between oil price volatility and stock returns in the Asia Pacific region of the world. Except for the countries of Indonesia, Singapore, and Sri Lanka, oil price volatility has little impact on the equity markets of the Asia Pacific region. Moreover, this result was detected after allowing for asymmetric oil price volatility effects. Evidence of an asymmetrical relationship between oil price volatility and stock returns is found only in the case of Singapore and this result is robust to two different measures of the exchange rage variable. These results suggest that, for the majority of countries studied, oil price volatility is not an important risk factor in explaining equity prices in Asia Pacific stock markets.

References

Adler, M, and Dumas, B. (1983). International portfolio selection and corporation finance: A synthesis, *Journal of Finance*, 38, 925-984.

Balke, N.S., Brown, S.P.A. and Yucel, M.K. (2002). Oil price shocks and the U.S. economy: Where does the asymmetry originate? *Energy Journal*, 23(3), 27-52.

Basher, S.A. and Sadorsky, P. (2006). Oil price risk and emerging stock markets, *Global Finance Journal*, 17, 2, 224-251.

Bollerslev , T., Chou, R.Y. and Kroner, K.F. (1992). ARCH modelling in finance: A review of the theory and empirical evidence *Journal of Econometrics*, 52, 5 – 59.

BP Statistical Review of World Energy, June 2006 (www.BP.com).

Chen, N.-F., Roll, R. and Ross, S.A. (1986). Economic forces and the stock market, *Journal of Business*, 59, 383-403.

Cunado, J., and Perez de Garcia, F. (2003). Do oil price shocks matter? Evidence for some European countries, *Energy Economics*, 25, 137-154.

Davis, S.J. and Haltiwanger, J. (2001). Sectoral job creation and destruction response to oil price changes, *Journal of Monetary Economics*, 48, 465-512.

Hamilton, J.D. and Herrera, A.M. (2004). Oil shocks and aggregate macroeconomic behavior: the role of monetary policy, *Journal of Money, Credit. and Banking*, 36(2), 265-286.

Engle, R.F. and Ng, V.K. (1993). Measuring and testing the impact of news on volatility. *Journal of Finance*, 48, 1749 - 1778.

Faff, R.W. and Brailsford, T.J. (1999). Oil price risk and the Australian stock market, *Journal of Energy Finance and Development*, 4, 69-87.

Ferson, W. W. and Harvey, C.R. (1994). Sources of risk and expected returns in global equity markets, *Journal of Banking and Finance*, 18, 775-803.

Ferson, W. W. and Harvey, C.R. (1995). Predictability and time-varying risk in world equity markets, *Research in Finance*, 13, 25-88.

Hammoudeh, S. and Aleisa, E. (2004). Dynamic relationships among GCC stock markets and NYMEX oil futures, *Contemporary Economic Policy*, 22, 2, 250-269.

Hammoudeh, S. and Choi, K. (2006). Behaviour of GCC stock markets and impacts of US oil and financial markets, *Research in International Business and Finance*, 20, 22-44.

Hooker, M.A. (1996). What happened to the oil price macroeconomy relationship? *Journal of Monetary Economics*, 38, 195-213.

Hooker, M.A. (2002). Are oil shocks inflationary? Asymmetric and nonlinear specifications versus changes in regime, *Journal of Money, Credit, and Banking*, 34, 540-561.

Huang, R. D., Masulis, R. W. and Stoll, H.R. (1996). Energy shocks and financial markets, *Journal of Futures Markets*, 16, 1, 1-27.

Jones, C.M. and Kaul, G. (1996). Oil and the stock markets, *Journal of Finance* ,51, 2, 463-91.

Jones, D.W., Leiby, P.N., and Paik, I.K. (2004). Oil price shocks and the macroeconomy: What has been learned since 1986, *The Energy Journal*, 25, 2, 1-32.

Kaneko, T. and Lee, B.S. (1995). Relative importance of economic factors in the U.S and Japanese stock markets, *Journal of the Japanese and International Economies*, 9, 290-307.

Lee, B.R. Lee, K. and Ratti, R.A. (2001). Monetary policy, oil price shocks, and the Japanese economy, *Japan and the World Economy*, 13, 321-349.

Mork, A. (1989). Oil and the macroeconomy when prices go up and go down: An extension of Hamilton's results, *Journal of Political Economy*, 97, 740-744.

Nandha, M. and Hammoudeh, S. (2007). Systematic risk, and oil price and exchange rate sensitivities in Asia-Pacific stock markets, *Research in International Business and Finance*, 21, 326-341.

Nelson, D.B. (1991). Conditional heteroskedasticity in asset returns: A new approach. *Econometrica*, 59, 347-70.

Papapetrou, E. (2001). Oil price shocks, stock markets, economic activity and employment in Greece, *Energy Economics*, 23, 511-532.

Ross, S. (1989). Information and volatility: The no-arbitrage martingale approach to timing and resolution irrelevance, *Journal of Finance*, 44, 1-17.

Sadorsky, P. (1999). Oil price shocks and stock market activity, *Energy Economics*, 21, 5, 449-469.

Sadorsky, P. (2003). The macroeconomic determinants of technology stock price volatility, *Review of Financial Economics*, 12, 191-205.

Sadorsky, P. (2004). Stock markets and energy prices, *Encyclopedia of Energy*, Volume 5, 707-717. Elsevier, New York.

The Economist (2006). *World in Figures*, Profile Books, London.

In: New Research on Energy Economics ISBN 978-1-60456-354-2
Editors: P.G. Caldwell and E.V. Taylor, pp. 135-148 © 2008 Nova Science Publishers, Inc.

Chapter 4

CLIMATE CHANGE AND TECHNOLOGY POLICY: DOUBLE MARKET FAILURE AND THE NEED FOR POLICY INTEGRATION

Mitsutsugu Hamamoto *
Faculty of Economics, Dokkyo University
Address: 1-1 Gakuen-cho, Soka-shi, Saitama 340-0042, Japan

Abstract

Existing literature on environmental economics has mainly focused on comparative analysis of environmental policy instruments such as direct regulations, emission taxes and emissions trading. In the literature, market-based approaches are recommended as policy instruments for combating climate change. However, new technologies must be created for the long-term solution to climate change problem. Although environmental policy can treat external diseconomy, market failure associated with research and development (R&D) activities must be addressed by technology policy: we are faced with the double market failure. Therefore, the integration of environmental and technology policies is needed for combating climate change. Japan's experience of pollution control in the 1960s and 1970s provides a case of such policy integration. This chapter investigates whether Japan's policy responses to air pollution during the period were effective. Recently, in Japan, a new approach was adopted in order to enhance the energy efficiency of appliances and vehicles. This is called the Top Runner Program, which may be able to promote R&D competition among producers. The chapter also discusses the effectiveness of the program.

1. Introduction

Comparative analysis of market-based and command-and-control approaches has been one of the main issues in the literature on environmental economics. Theoretical studies support the assertion that market-based instruments such as emission taxes and tradable permits can achieve environmental targets more efficiently than performance and technology standards.

*E-mail address: hmitutug@dokkyo.ac.jp; Phone/Fax: +81-48-942-6465

In addition, economic analyses generally conclude that market-based approaches are theoretically more attractive in incentives for research and development (R&D) than command-and-control approaches. While most researchers compare a single instrument with another one and assert the advantage of market-based approaches, there are many experiences of environmental policies combining several instruments. OECD (2006) states that environmentally related taxes are often used in combination with one or more other instruments such as direct regulations, subsidies, labeling systems, and voluntary agreements. There are many cases in which the revenues of environmentally related taxes in OECD countries are earmarked for environmental purposes. The Dutch system of water management, which includes wastewater levies, uses the revenues from the levies for pollution control (Andersen, 1999). In Sweden, a charge on nitrogen oxides (NO_x) emissions from energy generation at combustion plants was introduced in 1992. Revenues raised by the charge are refunded to the plants in proportion to their production of useful energy. This charge system can provide the plants with an incentive to reduce NO_x emissions per unit of energy (OECD, 2007). In addition, there can be cases in which emissions trading systems work with command-and-control approaches. In the United States, the sulfur dioxide allowance trading program was introduced in the 1990s in order to combat acid rain, while sulfur oxide pollution has been controlled by the direct regulations designed to achieve ambient standards to protect local air quality (Tietenberg, 1995).

In the European Union (EU), concern about the integration of environmental and other relevant policies has been growing. The Environmental Action Programs of the EU has claimed that environmental policy integration is needed in several fields such as agriculture, energy, industry, transport, and tourism. The aim of such policy integration is to change the nature of economic and other social activities in order that it may be consistent with the concept of sustainable development. However, it is pointed out that the EU and the Member States have been faced with bottlenecks that obstruct progress toward achieving environmental policy integration (Lenschow, 2002).

We lack experience of the full implementation of environmental policy integration. Under such a situation, investigations into several cases of public policies that can be interpreted as examples of environmental policy integration would be useful for understanding whether the integration of environmental and other policies is effective. Several European countries have introduced carbon dioxide (CO_2) taxes and used the revenues from the taxes to reduce taxes on labor. This can be considered as the integration of environmental policy with policy for reforming the tax system. Japan's experience of pollution control during the 1960s and 1970s provides a case of the integration of environmental and technology policies.

This chapter attempts to establish categories of combinations of multiple policies for environmental protection and discuss the application of the policy mixes in the categories to strategies for combating climate change. Section 2 provides three categories of combinations of multiple policies: instrument mixes, policy integration for economy-wide efficiency, and policy integration for technological innovation. The chapter focuses on the policy integration for technological innovation and explores Japan's public policies for pollution control and energy efficiency enhancement in order to derive lessons for the design of policy integration for reducing greenhouse gas (GHG) emissions. Section 3 investigates whether Japan's policy responses to air pollution during the 1960s and 1970s were effective.

Recently, in Japan, a new approach was adopted in order to enhance the energy efficiency of appliances and vehicles. This is called the Top Runner Program. Section 4 discusses the effectiveness of the program. Section 5 concludes the chapter with some implications for climate policy.

2. Combining Multiple Policies for Climate Change

2.1. Instrument Mixes

Economists have discussed the efficiency of market-based approaches under uncertainty. Economic analysis of climate change suggests that the marginal cost curve for reducing GHG emissions is very steep, while the marginal benefit curve for reducing emissions is fairly flat because the damages by climate change result from the overall stock of GHGs in the atmosphere, which is the accumulation of emissions over many years. In this situation, welfare loss is smaller under a tax policy than under a tradable permit system when the marginal costs are uncertain (Weitzman, 1974). Therefore, economic theory suggests that price approaches such as taxes would be more efficient than quantity approaches like emissions trading systems for controlling GHG emissions. However, a tax policy makes firms bear not only abatement costs but also tax payments for residual emissions, which would bring about large transfers of income from firms to the government. This is the main reason why firms oppose emission taxes. Thus, the introduction of a tax for reducing carbon emissions is likely to be faced with political difficulties (McKibbin and Wilcoxen, 2002).

Several countries have adopted tax policies that are combined with voluntary agreements. The CO_2 tax policy in Denmark has a scheme in which firms entering into an agreement on energy efficiency improvement with the Danish Energy Agency receive rebates on their CO_2 tax payments. In 2001, the Climate Change Levy was introduced in the United Kingdom. Under this levy system, energy-intensive sectors can obtain an 80% reduction in the tax rate if they enter into the Climate Change Agreements on the targets for carbon emission reductions or energy efficiency improvements (OECD, 2006). Tax policies combined with voluntary agreements like these cases make it possible to mitigate distributional effects on firms and adverse impacts on their international competitiveness that are caused by the introduction of taxes for reducing carbon emissions. However, the cost-efficient attainment of emission targets is unlikely to be realized under these policy schemes because the marginal abatement costs of regulated entities cannot be equalized in general.

Inefficiency under uncertainty and large transfers of income are the serious disadvantages of tradable permits and taxes, respectively, when these instruments are used alone to reduce GHG emissions. To overcome these problems, a hybrid approach is proposed (e.g., McKibbin and Wilcoxen, 2002): a tradable permit system combined with a tax that functions as a cap on permit prices. The aim of this approach is to prevent permit prices from soaring upward when abatement costs turn out to be much higher than expected. If the market prices of permits reach a specified level (known as a trigger price), firms can purchase additional permits at the price from the government. This system is called a "safety valve." Under the hybrid approach, firms can avoid unexpectedly high compliance costs, while a given emission target is effectively abandoned when a trigger price comes into effect.

The Regional Greenhouse Gas Initiative (RGGI), an agreement on climate change policy measures among the northeastern states in the United States, includes the establishment of a cap-and-trade system with a safety valve that is different from the above-mentioned one—if permit prices reach a trigger level for a sustained period, the compliance period will be extended. Furthermore, when this cannot be effective in reducing permit prices, regulated firms will be allowed to use offsets (i.e., conducting GHG reduction projects in and outside the RGGI member states) for up to 20 percent of their emissions.

2.2. Policy Integration for Economy-Wide Efficiency

Recently numerous studies on the relationship between environmental policy for reducing CO_2 emissions and the tax system have emerged. Some environmental economists suggest that carbon taxes can contribute not only to mitigating global warming but also to improving the efficiency of the tax system if the carbon tax revenues are used to lower existing distortionary taxes such as capital and labor income taxes. This win-win property of carbon taxes is called the strong double dividend. While economists generally agree that revenue recycling per se reduces the net cost of a carbon tax (a weak version of the double dividend), the strong double dividend hypothesis, which asserts that introducing a carbon tax accompanied by revenue recycling could generate net economic gains for society in addition to the benefits from CO_2 emission reductions, has elicited discussions on its validity.

Theoretical studies demonstrate that the strong version of the double dividend hypothesis fails to be supported. A carbon tax itself is distortionary because it increases the costs to firms of producing output, which will lead to slight reductions in the overall level of investment and employment. This is called the tax interaction effect. If the carbon tax revenues are used to reduce other distortionary taxes, welfare gains will be generated: this is the revenue-recycling effect. Most studies using analytical and computable models show that the tax interaction effect outweighs the revenue-recycling effect (Bovenberg and de Mooij, 1994; Parry, 1995; Goulder, et al., 1998). Thus, in theory, the net impact of a carbon tax is to reduce the level of investment and employment in the economy. This result does not absolutely deny the merit of the integration of environmental policy and reforming the tax system. Non-revenue-raising instruments such as emission standards and freely allocated tradable permits cannot have the revenue-recycling effect. It could be concluded that the economic burden which carbon abatement policies impose on society can be mitigated if revenue-raising instruments (e.g., taxes and auctioned permits) are adopted and the revenues are used to lower other distortionary taxes, although the strong double dividend cannot be supported in general.

During the 1990s, several European countries (Finland, Sweden, Denmark, and the Netherlands) introduced CO_2 taxes accompanied by revenue recycling to reduce taxes on labor. Later, such environmental tax reforms were carried out in Germany and the United Kingdom. A research project coordinated by the National Environmental Research Institute, University of Aarhus, has analyzed the effect of the environmental tax reforms on economic growth in these countries, finding that over the last 17 years the tax reforms have contributed to both economic growth and reductions in CO_2 emissions.[1] This result suggests that the strong double dividend hypothesis cannot be rejected in practice, which is

[1]See the "National Environmental Research Institute, University of Aarhus" website at http://www.dmu.dk/International/News/CO2tax.htm (accessed October 12, 2007).

contrary to the theoretical conclusion mentioned above. To reconcile this contradiction, further investigations will be needed in both theoretical and empirical studies of environmental tax reform.

2.3. Policy Integration for Technological Innovation

The approach to analyzing R&D incentives with which environmental policy instruments provide firms is traditionally to make a comparison of cost savings by developing and adopting new technologies under various instruments. Using this approach, Milliman and Prince (1989) conduct a comprehensive analysis of the relationship between environmental policy instruments and R&D incentives and find that taxes and auctioned permits provide the most incentives and emission standards provide the least. Montero (2002), pointing out that the theoretical studies using the above approach completely abstract from output market, compares R&D incentives offered by market-based instruments (auctioned and tradable permits) and command-and-control approaches under the conditions of oligopolistic output and/or permit markets. He shows the possibility that command-and-control approaches provide greater R&D incentives than market-based instruments if R&D investment by a firm affects its rival's cost through the output and permit markets, although this result depends on the specification of production cost functions (Bruneau, 2004).

The theoretical basis for introducing environmental policy instruments is the existence of external diseconomy that causes welfare losses for the economy. Technological innovation induced by environmental policy instruments is, so to speak, the ancillary benefit of introducing the instruments. Considering the economics of technological change, we are faced with another market failure. Knowledge production is mainly characterized by uncertainty and appropriability problem. These characteristics are the generic sources of market failure. Uncertainty associated with research activity makes it difficult to finance R&D investment through capital market mechanisms. New knowledge created by an innovator can bring spillover benefits to other parties without compensating the innovator. This impairs the ability of innovators to realize a reasonable rate of return on their R&D activities. Thus, uncertainty and appropriability problem lead to underinvestment in R&D. Market incentive mechanisms are likely to fail to produce the socially optimal rate and direction of innovative activity (Geroski, 1995). On the other hand, there are theoretical studies emphasizing the possibility that private rivalry for the rewards of a patent may lead to excessive investment in R&D. However, empirical studies show that the private rate of return on R&D is well below the social rate of return (Griliches, 1992), suggesting a tendency toward underinvestment in R&D. Market failure associated with R&D activities must be addressed by technology policy such as subsidizing R&D expenditures, cooperative R&D ventures, and the institutional design of patent.

In the long run the development of new technologies that can contribute to stabilizing atmospheric concentrations of GHGs in an economically feasible manner is essential for combating climate change. When pollution control technologies are underdeveloped and considerable costs are needed for abatement, establishing strict environmental regulations will be politically difficult. In this situation, a regulatory agency must set and strengthen regulations in terms of whether pollution control technologies that can reduce emissions at reasonable cost are available. If new technologies that can lower abatement costs are developed and diffused, they will contribute not only to attaining given levels of total emissions at lower cost, but also to making it possible to strengthen emission regulations in order to achieve more desirable levels of environmental targets. In other words, such new technologies may bring additional environmental benefits if proper policy responses are implemented. Thus, R&D investments in environmental technologies will produce social returns that consist of abatement cost reductions and environmental benefits. Such social returns on environmental R&D investments will exceed the private returns. Therefore, the integration of environmental and technology policies is needed under the situation where techniques for controlling pollution are immature.

Grubb and Ulph (2002) stress that it is necessary to understand better what designs of combinations of environmental and technology policies are effective, while there is very little literature in this research field.[2] Japan's experience of pollution control during the 1960s and 1970s, when pollution abatement techniques were immature, provides a case of the integration of environmental and technology policies. In the next section, Japan's public policies for controlling air pollution are discussed in order to derive lessons for the design of policy integration for reducing GHG emissions.

3. Japan's Public Policies for Controlling Air Pollution

Japan's public policies for controlling sulfur oxides (SO_x) and NO_x emissions during the 1960s and 1970s are characterized as the integration of command-and-control approaches to improving air quality and R&D subsidies for the development of pollution abatement technologies. The regulatory agency established environmental targets for atmospheric concentrations of SO_x and NO_x. To achieve these targets, emission standards applied to individual installations were set and strengthened gradually in terms of whether pollution abatement technologies that can reduce emissions at reasonable cost are available. Subsidies for R&D investment in pollution abatement technologies included grants from the central government and low-interest loans made by public financial institutions. Firms in the steel and automotive parts industries organized technology research associations in order to develop pollution control technologies under the national system of the promotion of cooperative R&D activity. In addition, a government-sponsored research project was conducted for developing flue-gas desulfurization technology. To assist polluting firms to invest in pollution control, public financial institutions made low-interest loans and tax incentives were offered. Such subsidies for pollution control investment may have contributed to the diffusion of newly developed pollution control technologies.

To illustrate the efficacy of the above policy integration, a simple model is presented. Suppose that a regulatory agency strengthens an emission standard if a new abatement technology is successfully developed. The current level of the emission standard and the level of stricter standard after succeeding in developing a new technology are represented as s_0 and s_1, respectively. The stricter standard brings environmental benefits denoted by $E[s_1]$, and $E[s_0]$ shows environmental benefits brought by the current standard ($E[s_1] - E[s_0] > 0$).

Consider a representative firm that invests in R&D activity to lower its own cost of

[2]For theoretical analysis of combining emission taxes and technology policy, see Katsoulacos and Xepapadeas (1996) and Katsoulacos, et al. (2001).

pollution abatement under the emission standard. $R(\cdot)$ represents the R&D cost that is an increasing function of R&D investment level e ($R' > 0$ and $R'' > 0$). $P(\cdot)$ is the probability of success in the development of a new abatement technology, which increases with R&D investment ($P' > 0$, $P'' < 0$, $P(0) = 0$, and $\lim_{e \to \infty} P(e) = 1$).

Under the current emission standard, a profit earned by the firm using an old abatement technology before deduction of the R&D cost is denoted by $\pi[C_0, s_0]$, where C_0 is the cost of abatement using the old technology. When a new abatement technology is successfully developed and the emission standard is strengthened, the firm, using the new technology, can earn a profit of $\pi[C_1, s_1]$ (before deduction of the R&D cost), where C_1 shows the cost of abatement using the new technology. While the new abatement technology can lower the abatement cost, the stricter standard leads to an increase in the cost. Therefore, the relationship between $\pi[C_0, s_0]$ and $\pi[C_1, s_1]$ has two cases: (a) $\pi[C_0, s_0] < \pi[C_1, s_1]$, and (b) $\pi[C_0, s_0] \geq \pi[C_1, s_1]$.

The regulatory agency solves the optimization problem as follows:

$$\max_{e} P(e)\{E[s_1] + \pi[C_1, s_1]\} + (1 - P(e))\{E[s_0] + \pi[C_0, s_0]\} - R(e). \quad (1)$$

The first order condition is given by:

$$P'\{E[s_1] - E[s_0] + \pi[C_1, s_1] - \pi[C_0, s_0]\} = R'. \quad (2)$$

On the other hand, since the firm cannot receive the additional environmental benefits brought by the stricter standard (i.e., $E[s_1] - E[s_0]$), the first order condition of the profit maximization problem for the firm is:

$$P'\{\pi[C_1, s_1] - \pi[C_0, s_0]\} = R'. \quad (3)$$

When $\pi[C_0, s_0] < \pi[C_1, s_1]$, Equation (3) shows that the firm invests in R&D activity to develop the new technology that can lower its own abatement cost. However, the level of R&D investment determined by (3) is not socially optimal. Figure 1 illustrates the discrepancy between the socially and privately optimal levels of R&D investment. The upward sloping curve labeled MRDC depicts the marginal R&D cost (i.e., R'). Two downward sloping curves labeled MPB and MSB are the marginal private benefit of R&D investment (i.e., $P'\{\pi[C_1, s_1] - \pi[C_0, s_0]\}$) and the marginal social benefit (i.e., $P'\{E[s_1] - E[s_0] + \pi[C_1, s_1] - \pi[C_0, s_0]\}$), respectively. The socially optimal level of R&D investment is e^* and the privately optimal level e^p. The figure shows that $e^* > e^p$: the R&D investment level that the firm privately chooses is too low in comparison to the socially efficient level. This provides a theoretical foundation for introducing technology policy combined with command-and-control approaches to environmental policy.

When $\pi[C_0, s_0] \geq \pi[C_1, s_1]$, the firm has no incentive to be engaged in R&D activity because it cannot produce the private return on R&D investment. However, if $P'\{E[s_1] - E[s_0] + \pi[C_1, s_1] - \pi[C_0, s_0]\} > 0$, it is theoretically justifiable that resources in the economy are allocated to R&D activity. In this case, public policies such as R&D subsidies or government-sponsored research projects will be needed in order to encourage R&D investment.

Thus, technology policy can play a role in the promotion of R&D activity when a regulatory agency strengthens emission standards in terms of whether private firms succeed in

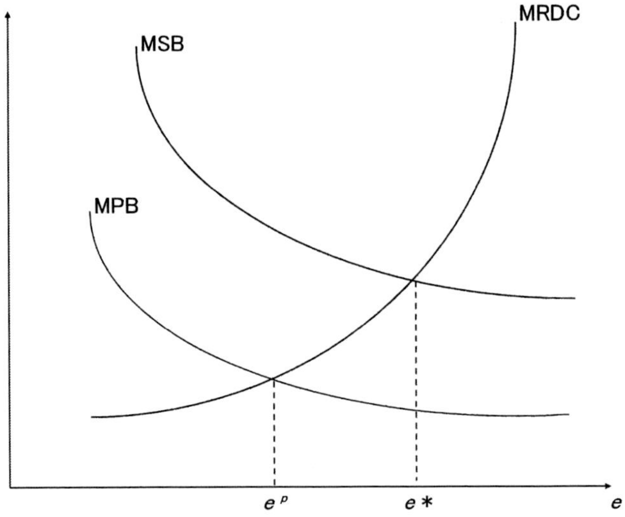

Figure 1. Socially and Privately Optimal Levels of R&D Investment.

developing new abatement technologies that can reduce emissions at reasonable cost. The above model does not treat market failure caused by uncertainty and spillover effect associated with knowledge production. Consideration of such market failure will reinforce the necessity for technology policy combined with environmental policy.

There are several reviews of Japan's pollution control policies in the 1960s and 1970s. Weidner (1995) claims that environmental policy for controlling SO_x and NO_x emissions from stationary sources in Japanese industries such as steel, paper and pulp and chemicals can be evaluated as a successful case. He points out that such policy had no adverse impact on the Japanese macro-economy. OECD (1977) states that, although pollution control pushed up the production costs of Japanese industries, the additional costs did not seriously impair their competitiveness. These suggest that Japan's style of the integration of environmental and technology policies could mitigate the adverse effects of pollution control activities on the domestic economy.

The above discussion does not insist that environmental regulations play only a subordinate role in providing firms with R&D incentives. Rather, there is empirical evidence that the command-and-control approaches adopted in Japan had a great impact on the development of pollution abatement technologies. Hamamoto (2006) examines the effect of the stringency of environmental regulations (as measured by pollution control expenditures) on R&D activity (as measured by R&D expenditures) in Japanese manufacturing industries, finding a significant positive relationship between the pollution control expenditures and the R&D expenditures. This indicates that the pressure of environmental regulations would spur innovative activity.

Hamamoto (2006) also estimates the impact of increases in R&D investment on the total factor productivity (TFP) of the industries, showing that increases in R&D investment stimulated by regulatory stringency have a significant positive effect on the TFP growth rate and produce a high rate of return. This implies that the burden of environmental compliance would provide polluting firms faced with both environmental and financial constraints with an incentive to develop cleaner production technologies.

Several examples of cleaner production technologies that were developed in Japanese heavy industries faced with strict environmental regulations can be cited. In the pulp and paper industry, wastewater containing a chemical material used for producing pulp from woodchips is recovered from pulp mills. The chemical material and a source of energy supply can be obtained by burning the wastewater. This recovery process contributes not only to a reduction in water pollution but also to an improvement in productivity (Nakanishi, 1994). In the chemical products industry, mercury emissions from caustic sodium plants were reduced by converting production process using mercury cells into diaphragm cell technology in the 1970s. However, the diaphragm process brought about a deterioration in the quality of products and an increase in production costs, which gave firms in the caustic sodium industry an incentive to devote their efforts to the development of a mercury-free, efficient and quality-improving production technology. Thus, ion exchange membrane process was developed and contributed to the amelioration of profitability in the 1980s (Nippon Soda Kogyokai, 1982). In the iron and steel industry, several energy conservation technologies (e.g., a computerized system of operating thermal devices, the elimination and concatenation of production processes, and waste heat recovery) contributed both to reductions in air pollution and to productivity improvements (Committee on Japan's Experience in the Battle against Air Pollution, 1997).

If global and domestic climate policies establish targets for controlling the total volume of GHG emissions, the targets must be strengthened through several (or many) phases in order to achieve the stabilization of atmospheric concentrations of GHGs. When a domestic regulatory agency strengthens the target for total GHG emissions, it may be faced with political difficulties under the situation where firms lack abatement technologies that can reduce GHG emissions at reasonable cost. Japan's style of the integration of environmental and technology policies will make it possible to avoid such difficulties and mitigate the serious adverse effects of GHG abatement activities on the economy.

However, technology policy accompanied by government expenditure, such as R&D subsidies, needs financial resources. If the government expenditure is financed through increases in commodity or income taxes, welfare losses are generated because of distorting consumer choices or dampening work effort and investment. When such welfare losses outweigh the social net benefit brought by the integration of environmental and technology policies, such policy integration may not be justifiable. Thus, the use of R&D subsidies should be limited to financial assistance to research activities with high-spillovers or serious difficulty in financing. In addition to such R&D subsidies, the government must provide institutional settings that can promote R&D competition among private firms for the development of low carbon technologies. The Top Runner Program, which has recently been implemented in Japan in order to enhance the energy efficiency of appliances and vehicles, may be an example of such institutional settings. The next section discusses the effectiveness of the program.

4. New Approach to Enhancement of Energy Efficiency

In April 1999, the Top Runner Program, the aim of which is to reduce energy consumption in the household and transportation sectors, was introduced by a revision of the Law concerning the Rational Use of Energy. The program requires manufacturers of energy-using products to meet the Top Runner standards, which are future energy efficiency requirements based on the best performance of current technologies. The standards are applied to selected groups of energy-using products. At the start of the program, nine groups of appliances and vehicles such as air conditioners, cathode-ray tube television sets, video cassette recorders, and passenger vehicles were selected. As of April 2007, the product groups to which the Top Runner standards are applied are as follows: gasoline/diesel/LPG passenger vehicles, air conditioners, fluorescent lights, cathode-ray tube/liquid crystal display/plasma television sets, copying machines, computers, magnetic disk units, diesel/gasoline freight vehicles, video cassette recorders, DVD recorders, electric refrigerators, electric freezers, gas/oil space heaters, gas cooking appliances, gas water heaters, oil water heaters, electric toilet seats, vending machines, oil-filled/molded transformers, electric rice cookers, and microwave ovens. Under this program, the regulatory agency uses the highest efficiency of an energy-using product achieved by a manufacturer as the basis of a new energy efficiency standard for the product.[3] Other firms in the market are required to meet the new (Top Runner) standard within a certain period. More specifically, each firm must make sure that the weighted average of energy efficiency of products included in a category the program designates meets the standard within a fixed time limit. Thus, the Top Runner Program can provide firms with incentives to develop more energy efficient products.[4] In addition, the program may have an effect on R&D competition: it is likely that firms will invest more in their R&D activities to be a "top runner"—the first to achieve the highest energy efficiency if R&D spillovers do not seriously undermine the appropriability of returns on inventive activity and it is burdensome for manufacturers except the top runner to achieve the highest efficiency established as a mandatory standard. The program might be able to provide manufacturers with more powerful incentives to develop energy efficient products as compared with existing approaches such as minimum and average energy standards.

In order to promote the diffusion of highly efficient products, consumers' awareness of their own benefits from using them is needed. In 2000, the energy conservation labeling system for several categories of appliances to which the Top Runner standards are applied was launched. This labeling system provides consumers with information about to what extent appliances have achieved the standards.[5] A similar system of labeling for the energy efficiency of vehicles was introduced in 2004. Thus, the Top Runner Program is working with the labeling systems.

To investigate whether and to what extent the Top Runner Program has had effects on the R&D competition among the manufacturers and the energy efficiency of appliances and vehicles, empirical research using econometric methods needs to be carried out. Although such research remains as a future task, there are statistical data suggesting how effective the Top Runner Program has been in enhancing the energy efficiency of the products. Figure 2 illustrates the trend in the electricity consumption by air conditioners. The average annual

[3]The Energy Efficiency Standards Subcommittee, established under the Advisory Committee for Natural Resources and Energy (an advisory body to the Minister of Economy, Trade and Industry), deliberates and makes decisions on the Top Runner standard setting.

[4]More detailed information on the Top Runner Program can be found on the "Energy Conservation Center, Japan" (ECCJ) website at http://www.eccj.or.jp/top_runner/index_contents_e.html (accessed April 26, 2007).

[5]For example, if a product is 20% more (less) efficient than the Top Runner standard, it has a label indicating "120% (80%)".

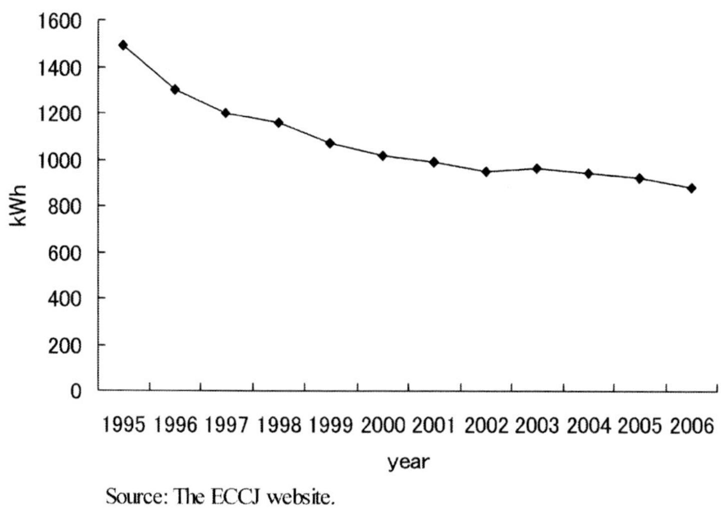

Figure 2. Annual Electricity Consumption of Air Conditioners.

electricity consumption of air conditioners was 1,068 kWh in 1999 and was cut down to 882 kWh in 2006: after the introduction of the Top Runner Program, a 17.4% reduction in electricity consumption was realized. Figure 3 shows the trend in the electricity consumption of cathode-ray tube televisions (CRT-TVs). The average electric power consumption of CRT-TVs in active mode decreased gradually, and so did the average annual electricity consumption. While the average annual electricity consumption of the 1999 models is 201kWh, the 2006 models, on average, use 144kWh—a 28.4% reduction compared with the 1999 models.

These figures do not necessarily imply that the Top Runner Program is the most effective factor in such reductions in electricity consumption. It will be worth investigating to what extent the program has spurred R&D activities and eventually enhanced the energy efficiency of appliances and vehicles.

5. Conclusion

What is essential for the long-term solution to climate change is to achieve the transition from carbon intensive technologies (i.e., technologies heavily dependent on fossil fuels) to low carbon ones. Such transition may unavoidably impose considerable burden on society because there will be serious barriers to changing the paths of technological development characterized by lock-in or path dependency. Policy choices associated with climate change problem determine the extent and distribution of the burden. To minimize the burden imposed by the transition of technological paths, investigations on the optimal integration of environmental and technology policies are needed. Technology policy should include

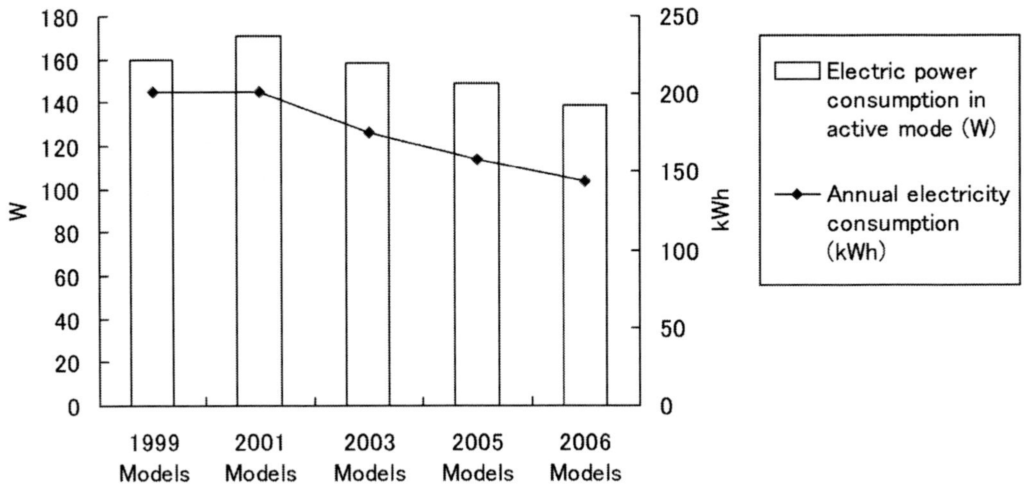

Source: The ECCJ website.

Figure 3. Annual Electricity Consumption of CRT-TVs

the promotion of R&D competition as well as R&D subsidies in order to minimize welfare losses caused by financing such subsidies through taxation. The Top Runner Program might function as a competition policy aimed at the creation of low carbon technologies. While environmental policy plays a significant role in sending a clear signal that induces R&D resources to be devoted to creating environment-friendly technologies, it should be primarily designed so as to achieve cost-efficient reductions in GHG emissions (i.e., the equalization of the marginal abatement costs of emission sources). Although a regulatory agency may seek to use a tougher environmental policy (a stricter emission limit or a higher emission tax) in order to spur technological development, introducing such a policy is likely to be faced with political difficulties. When quantitative limits on GHG emissions are used to address climate change, the integration of environmental and technology policies will make it politically feasible to strengthen the limits.

There has been much debate concerning the regulatory choice between price and quantity approaches in designing climate policy, while we have observed the recent movement toward widespread acceptance and application of cap-and-trade programs for controlling GHG emissions. Nordhaus (2007) criticizes quantity approaches to slowing global warming, like the Kyoto Protocol, for their disadvantages (e.g., inefficiency under massive uncertainties, high volatility in the market price of carbon, and the generation of artificial scarcities to encourage rent-seeking behavior) and recommends price approaches such as carbon taxes that are steadily raised over time so as to reflect the increasing prospective damages from climate change. Such price approaches may have difficulty in increasing the carbon taxes when abatement technologies that can reduce GHGs emissions at reasonable cost are not available. Therefore, the integration of environmental policy using price approaches with technology policy would contribute to enhancing the political feasibility of raising carbon taxes.

References

[1] Andersen, M. S. (1999) Governance by green taxes: implementing clean water policies in Europe 1970–1990. *Environmental Economics and Policy Studies*, 2, 39–63.

[2] Bovenberg, A. L., and de Mooij, R.A. (1994) Environmental levies and distortionary taxation. *American Economic Review*, 84 (4), 1085–1089.

[3] Bruneau, J. F. (2004) A note on permits, standards, and technological innovation. *Journal of Environmental Economics and Management*, 48, 1192–1199.

[4] Committee on Japan's Experience in the Battle against Air Pollution (1997) *Japan's Experience in the Battle against Air Pollution: Working towards Sustainable Development.* Tokyo: Pollution-Related Health Damage Compensation and Prevention Association.

[5] Geroski, P. (1995) Markets for technology: knowledge, innovation and appropriability. In P. Stoneman (Ed.), *Handbook of the Economics of Innovation and Technological Change* (pp. 90–131). Oxford: Basil Blackwell.

[6] Goulder, L. H., Parry, I. W. H., Williams III, R. C., and Burtraw, D. (1998) The cost-effectiveness of alternative instruments for environmental protection in a second-best setting. Resources for the Future, *Discussion Paper* 98–22.

[7] Griliches, Z. (1992) The search for R&D spillovers. *Scandinavian Journal of Economics*, 94 (Supplement), S29–S47.

[8] Grubb, M., and Ulph, D. (2002) Energy, the environment, and innovation. *Oxford Review of Economic Policy*, 18 (1), 92–106.

[9] Hamamoto, M. (2006) Environmental regulation and the productivity of Japanese manufacturing industries. *Resource and Energy Economics*, 28 (4), 299–312.

[10] Katsoulacos, Y., Ulph, A., and Ulph, D. (2001) The effects of environmental policy on the performance of environmental research joint ventures. In C. Carraro, and G.E. Metcalf (Eds.), *Behavioral and Distributional Effects of Environmental Policy* (pp. 309–339). Chicago: The University of Chicago Press.

[11] Katsoulacos, Y., and Xepapadeas, A. (1996) Environmental innovation, spillovers and optimal policy rules. In C. Carraro, Y. Katsoulacos, and A. Xepapadeas (Eds.), *Environmental Policy and Market Structure* (pp.143–150). Dordrecht: Kluwer.

[12] Lenschow, A. (2002) Greening the European Union: an introduction. In A. Lenschow (Ed.), *Environmental Policy Integration: Greening Sectoral Policies in Europe* (pp. 3–21). London: Earthscan.

[13] McKibbin, W. J., and Wilcoxen, P. J. (2002) *Climate Change Policy after Kyoto: Blueprint for a Realistic Approach.* Washington, D.C.: Brookings Institution Press.

[14] Milliman, S. R., and Prince, R. (1989) Firm incentives to promote technological change in pollution control. *Journal of Environmental Economics and Management*, 17, 247–265.

[15] Montero, J.-P. (2002) Permits, standards, and technology innovation. *Journal of Environmental Economics and Management*, 44, 23–44.

[16] Nakanishi, J. (1994) *Mizu no Kankyo-Senryaku (Environmental Strategies for Water Quality Management).* Tokyo: Iwanami Shoten (in Japanese).

[17] Nippon Soda Kogyokai (1982) *Nippon Soda Kogyo Hyakunenshi (The History of the Japanese Caustic Soda Industry).* Tokyo: Nippon Soda Kogyokai (in Japanese).

[18] Nordhaus, W. D. (2007) To tax or not to tax: alternative approaches to slowing global warming. *Review of Environmental Economics and Policy*, 1 (1), 26–44.

[19] OECD (1977) *Environmental Policies in Japan.* Paris: OECD.

[20] OECD (2006) *The Political Economy of Environmentally Related Taxes.* Paris: OECD.

[21] OECD (2007) *Instrument Mixes for Environmental Policy.* Paris: OECD.

[22] Parry, I. W. H. (1995) Pollution taxes and revenue recycling. *Journal of Environmental Economics and Management*, 29, S-64-S-77.

[23] Tietenberg, T. (1995) Tradeable permits for pollution control when emission location matters: what have we learned? *Environmental and Resource Economics*, 5, 95–113.

[24] Weidner, H. (1995) Reduction in SO_2 and NO_2 emissions from stationary sources in Japan. In M. Jänicke, and H. Weidner (Eds.), *Successful Environmental Policy: A Critical Evaluation of 24 Cases* (pp. 146–172). Berlin: edition sigma.

[25] Weitzman, M. L. (1974) Prices vs. quantities. *Review of Economic Studies*, 41, 477–491.

In: New Research on Energy Economics
Editors: P.G. Caldwell and E.V. Taylor, pp. 149-168
ISBN: 978-1-60456-354-2

Chapter 5

A PRELIMINARY SURVEY OF THE ECONOMICAL VIABILITY OF LARGE-SCALE PHOTOBIOLOGICAL HYDROGEN PRODUCTION UTILIZING MARICULTURE-RAISED CYANOBACTERIA

Hidehiro Sakurai[1,3,*]***, Hajime Masukawa***[1,2] ***and Kazuhito Inoue***[1,2]
[1]Research Institute for Integrated Science and [2]Department of Biological Sciences, Kanagawa University, 2946 Tsuchiya, Hiratsuka, Kanagawa, 259-1293, Japan
[3] School of Education, Waseda University, Nishiwaseda 1, Shinjuku, Tokyo 169-8050, Japan

Abstract

This paper briefly examines the future prospects for the economical viability of large-scale renewable energy production using maricultured cyanobacteria. In order to reduce CO_2 emissions from burning fossil fuels in appreciable amounts, the replacement energy source will by necessity be substantial in scale. Solar energy is the most likely candidate because the amount of solar energy received on the Earth's surface is vast and exceeds the anthropogenic primary energy use by more than 6,000 times. Although solar energy is abundant, its economical utilization is not straightforward because the intensity received on the surface of the earth is relatively low. Current research and development efforts are focused on the production of biofuels as renewable, economical feasible energy sources from the land biomass. We propose, however, for reasons of scale and to minimize further environmental harm, that the utilization of the sea surface is a more viable alternative to land biomass exploitation. The sea surface area available for energy production far exceeds available cropland and use of the sea will not take valuable cropland out of food production. Our current R and D strategy utilizes photosynthesis and the nitrogenase enzyme of cyanobacteria. The biological basis of relevant energy metabolism in cyanobacteria is briefly described. A model for future H_2 production systems is presented, and a very rough trial calculation of the cost of photobiological H_2 production is made in the hope that it may help the readers

* E-mail address: sakurai@waseda.jp , Tel: +81-3-5286-1508, Fax: +81-3-3207-9694, Mailing address: Shakujiidai 4-1-3-201, Nerima, Tokyo 177-0045, Japan. Correspondence to: Hidehiro Sakurai.

recognize the possibilities of large-scale H_2 production and understand the need for the research and development.

KeyWords: cyanobacteria economical viability hydrogen production mariculture nitrogenase solar energy conversion.

1. Introduction

1.1. Our Need for Developing Large-Scale Renewable Energy Sources

1.1.1. Climate Change

At the onset of the first oil crisis in the fall of 1973, many renewable biological energy sources were considered as alternatives to fossil fuels (Gibbs et al., 1973; Lien and San Pietro, 1975). The crisis was short-lived, as the supply of oil by OPEC was largely uninterrupted, and investors as well as the oil-consuming public lost interest in exploring renewable energy sources during the subsequent two to three decades. More recently, there has been a renewed recognition of the need for the development of solar energy as a result of concerns about the looming shortage of oil supplies and more importantly the global climate change caused by burning fossil fuels.

Recognizing the environmental risks brought about by ever-increasing concentrations of atmospheric greenhouse gasses (GHGs), most notably CO_2, the United Nations adopted the Framework Convention on Climate Change (UNFCCC) in 1992, and the Convention entered into force in March 1994. In the Convention, the signatory nations stated their commitment to protect the climate for the benefit of present and future generations (UNFCCC, 2007). In spite of this and other efforts of the international community to cut CO_2 emissions, including adoption of the Kyoto Protocol, UNFCCC, there is still no established world-wide framework for stabilizing the atmospheric concentration of greenhouse gasses at the levels required for preventing further dangerous anthropogenic interference with the global climate system.

In April 2007, R.K. Pachauri, the chairman of the IPCC (Intergovernmental Panel on Climate Change) spoke at the Sessions of the Subsidiary Bodies, UNFCCC, where he stated that, "Most of the observed increase in globally averaged temperatures since the mid-20th century is very likely due to the observed increase in anthropogenic greenhouse gas concentrations" (IPCC, 2007a). Likewise, the 4th IPCC Assessment Report (IPCC, 2007b) states: "The primary source of the increased atmospheric concentration of carbon dioxide since the pre-industrial period results from fossil fuel use, with land-use change providing another significant but smaller contribution. Annual fossil carbon dioxide emissions increased from an average of 6.4 GtC (23.5 $GtCO_2$) per year in the 1990s to 7.2 GtC (26.4 $GtCO_2$) per year in 2000–2005. Carbon dioxide emissions associated with land-use change are estimated to be 1.6 GtC (5.9 $GtCO_2$) per year over the 1990s. The global atmospheric concentration of carbon dioxide has increased from a pre-industrial value of about 280 ppm to 379 ppm in 2005. The annual carbon dioxide concentration growth rate was larger during the last 10 years (1995–2005 average: 1.9 ppm per year), than it has been since the beginning of continuous direct atmospheric measurements (1960–2005 average: 1.4 ppm per year)"

1.1.2. Essential CO_2 Emission Reductions

The ICPP Report (IPCC, 2007c) predicts that in order to stabilize the atmospheric CO_2 concentration at around 560 ppm, a level about twice that of the pre-industrial value, we must begin reducing CO_2 emissions by 2010–2030. Although there is considerable uncertainty in the estimates, one scenario predicts that we must limit the CO_2 emission from at least 50% of the 2000 level by the end of the century (2100). Less optimistic estimates require limiting CO_2 emissions to 0% of the 2000 level or even more drastically, require removal of CO_2 from the atmosphere (negative emissions) using technologies such as carbon capture and storage. Even if we succeed in stabilizing the CO_2 concentration at about 560 ppm, the global mean temperature will increase above the pre-industrial mean temperature by 2.8–3.2 °C, at equilibrium.

The International Energy Agency (IEA, 2007b) predicts in a Reference Scenario of World Energy Outlook 2007, that global primary energy demand will increase by 55% between 2005 and 2030, at an average annual rate of 1.8% per year, in the absence of new government measures to alter underlying energy trends.

Although there has been a steadily growing consensus in the world community that CO_2 emissions must be reduced, there has yet to be commitment to a global strategy for attaining CO_2 emission reductions at any level.

1.2. Estimate of Potential of Renewable Energy Production From Land Biomass

Biomass products such as wood, grass, and foodstuffs are considered to be sources of renewable energy, because of their capacity to capture CO_2 by photosynthesis as they grow, even though they ultimately emit CO_2 when they are used for energy production. According to the IEA BIOENERGY Report (IEA, 2007a), current global energy supplies are dominated by fossil fuels (388 EJ (EJ: 10^{18} J) per year), with biomass providing about 45 ± 10 EJ. On average, biomass contributes less than 10% to the total energy supply in industrialized countries, but the proportion is as high as 20–30% in developing countries. If the world community achieves a large-scale deployment of bioenergy technologies by the year 2050, the average energy production potential for biomass is estimated to be from 40 to 1100 EJ. In this case, the lowest estimate is based on the most pessimistic scenario of no additional farmland being available for energy production and with the biomass coming solely from traditional crop residues. Under the most optimistic scenario, higher levels of energy production from biomass could be achieved under well-managed, intensive agriculture concentrated on the best quality soils. Necessary management practices would include energy farming on current agricultural land, biomass production on marginal lands, and energy capture from other agricultural sources including forest residues, manure, and organic waste. Overall, the estimated range of the plausible potential for biomass energy production is 200 – 400 EJ.

Even with state-of-the-art biomass energy conversion technologies, the potential of bioenergy seems to be limited by the availability of fertile land and water. Furthermore, the destruction of forests to attain more arable land will result in the emission of even more CO_2

as the photosynthetic, CO_2-capturing and storing capacity of the forest is lost and ultimately organic substances stored in the forest soils decompose.

Under realistic scenarios, we cannot expect great amounts of additional energy to be produced from crops grown on land, as the world's primary energy consumption is vast and amounts to about 20 times of the digestive energy available from food (see, Sakurai and Masukawa, 2007). If the entire amount of cropland in the U. S. (about 400 million hectares) were used to produce corn solely for ethanol, there would still not be enough fuel to fulfill the 1990 level of U.S. gasoline consumption (estimated to be less than 90% of gasoline by Kheshgi et al. (2000)). According to the forecast of an energy analyst (T. Ueda, Toyota Motors Co.), the percentage of energy available in bioethanol as a motor fuel (about 1.8%) will make it a challenging energy source to use to achieve the goal of meeting 10% of the world gasoline consumption by 2020 (presentation at the 2006 Annual Meeting of the Japan Institute of Energy, 2006 August). Considering that motor fuels comprise roughly a quarter of fossil fuel energy consumption, using motor fuel blends of gasoline with ethanol of 10% in energy (about 16% in volume) would result in only about a 2–3% reduction of the current level of world fossil fuel consumption.

2. Our Proposals for Large-Scale Photobiological H_2 Production

2.1. Outlines of our Proposal

We will first describe an outline of our proposal in this section, followed by more detailed descriptions in the following sections. We present here information needed to evaluate the prospects of photobiological H_2 production capitalizing on the nitrogenase enzyme of mariculture-raised cyanobacteria. In proposing this system, we do not intend to criticize other systems including the hydrogenase-based system.

We have proposed nitrogenase-based photobiological production of H_2 using maricultured cyanobacteria (also called blue-green algae or cyanophytes), because we believe such systems have the potential to produce renewable energy of a scale large enough to meet much of the world-wide energy demand (Sakurai and Masukawa, 2007). We have chosen H_2 as the product of solar energy conversion processes, because H_2 is considered to be an ideal fuel for the following reasons: 1) when used as a fuel, in contrast with oils and coals that emit SO_x, CO_2, and sometimes soot during combustion, H_2 does not pollute the environment as the by-product of its production is water, 2) in contrast with electricity generated by photovoltanics, H_2 is easily stored and transported, 3) H_2 is expected to be directly used as an energy source of fuel cells (Levin et al., 2004; Mertens and Liese, 2004) in the near future.

2.2. The Prospects of Economically-Feasible Large-Scale H_2 Production

Our proposal for promoting the R and D of mariculture-based economical large-scale solar energy conversion is based on the following premises (Table 1).

Table 1. Prospects of photobiological H_2 production

A. The amount of solar energy received on the Earth's surface is huge: exceeds the anthropogenic primary energy use by more than 6,000 times
B. Expected whole sale: around \$1.5 - 7.5 m^{-2} yr^{-1} (The total cost of production of bio-solar energy should meet these economical constraints) Assumptions 1) Solar energy received on the surface of the earth: about 1,500 kWh m^{-2} yr^{-1} for the middle latitude regions 2) The amount of renewable energy produced per unit area will be about 15 kWh m^{-2} yr^{-1} (at 1% bio-solar energy conversion efficiency) 3) The sale price of the renewables at 10 – 50 cents kWh^{-1}
C. Grounds for future economical production at 10 - 50 cents kWh^{-1}: 1) The solar energy conversion efficiencies of 0.25-0.3% into the final commodities (corn grains) in U.S.A.. (our target: 1% efficiency) 2) The price of corn (Chicago) has been in the range of 8 - 21 cents kg^{-1} or about 1.8 - 5.1 cents kWh^{-1} in the past ten years (1997-2006) (lower than our target price)

2.3. Scale of Production

In identifying photobiological H_2 production R and D strategies, we must first calculate the final production scale required for meaningful reduction of CO_2 emissions.

According to the report of the International Energy Agency (IEA, 2007a), the total current (2004) commercial energy use amounts to some 467 EJ per year, of which fossil fuels account for 388 EJ per year. The total solar radiation received on the earth's surface is about 2.7×10^6 EJ per year (see Sakurai and Masukawa, 2007) (the value of 3.9×10^6 EJ per year theoretical potential as described in the above report, possibly including energy absorbed by the atmosphere). In order to replace 10% of the 2004 level of commercial energy use (3.88 EJ) by H_2-based energy production systems, according to our proposal, the sea surface area required for the production is roughly calculated to be 1.1×10^6 km^2 (about the same size as the country of Bolivia, or about twice that of France, for example).

Assumptions used for the sea surface area calculation:

1) Total solar radiation: 2.7×10^6 EJ per year, which leads to 5.3×10^9 J m^{-2} $year^{-1}$ (1,680 kWh m^{-2} $year^{-1}$ or 14.5 MJ m^{-2} day^{-1}) on average.
2) Cyanobacteria photobiologically produce H_2 at 1% efficiency vs. total solar radiation (5.3×10^7 J m^{-2} $year^{-1}$).
3) Finally, available H_2 energy: two thirds of biologically produced H_2 energy, consuming one third needed for harvesting, purification, storage, and transportation to land (3.5×10^7 J m^{-2} $year^{-1}$, in the final product).
4) Area required: $38.8 \times 10^{18}/(3.5 \times 10^7) = 1.1 \times 10^{12}$ (m^2) ($=1.1 \times 10^6$ km^2)

The required sea surface is extremely vast. Considering that a crop farmer in the U.S.A. needs several hundred ha (several km^2) of farmland for living even with a subsidy by the government, the area used by any company or organization for photobiological H_2 production should be much larger than several km^2. If the size of the bioreactor is 25 m in width and 200 m in length (0.5 ha), for example, 200 bioreactors are required for 1 km^2, and 2.2×10^8 bioreactors in order to replace 10% of the 2004 level of commercial energy use.

2.4. Outline of a Plausible General Scheme for Photobiological H_2 Production and Utilization

Most of the technologies required for large-scale photobiological H_2 production by cyanobacteria and subsequent utilization of H_2 are still in the beginning stages of development. Acknowledging that there are a lot of uncertainties, we will briefly present a plausible general scheme for future production so that the readers may understand the research and development strategies needed to develop H_2-based energy production systems (Figure 1).

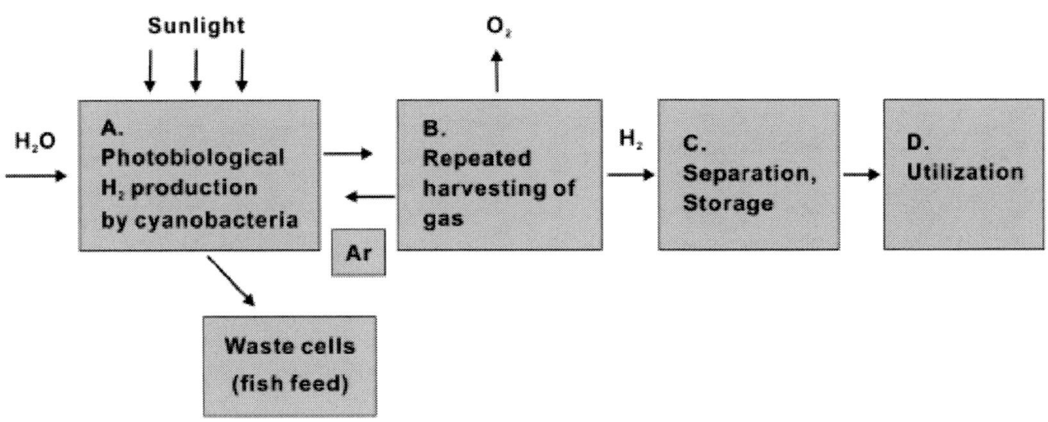

Figure 1. General scheme for photobiological H_2 production and utilization. The substrate (raw materials) of H_2 is water. CO_2 is required for growth and maintenance of the cyanobacteria cells and once the culture is established, its re-addition is unnecessary because it is recycled in the bioreactor.

Process A: Photobiological H_2 production by cyanobacteria

Some of cyanobacteria produce H_2 by photosynthetic activities with sunlight as the sole source of energy (see below). The substrate is water and products are H_2 and O_2. The cyanobacterial culture medium (fresh water or sea water) is simple in composition and contains mineral nutrients, and the gas phase in the bioreactor is Ar plus low concentrations of CO_2 and N_2. After the initial stage of cell growth, further supply of CO_2 is not necessary, because it is recycled in the bioreactor (Figure 2A).

The remaining biological waste (cells) can be recycled as fish feed (Murray and Mitsui, 1982).

Some areas of calm sea (such as inland seas) and ocean (the calm belts, e.g. the doldrums near the equator and the horse latitudes of about 30° north or south as exemplified by the

"Mysterious Bermuda triangle") seem to be especially suited for such large-scale mariculture in inexpensive plastic bags. From simulations of sea movement, observations of floating marine debris accumulation, and by taking into account the surface currents consisting of Stokes drift, Ekman drift, and geostrophic currents, Kubota et al. (2005) have shown that the debris tends to accumulate at about 30° north or south. In such regions, a great number of bioreactors can be simply floated on the calm water surface without need for mooring.

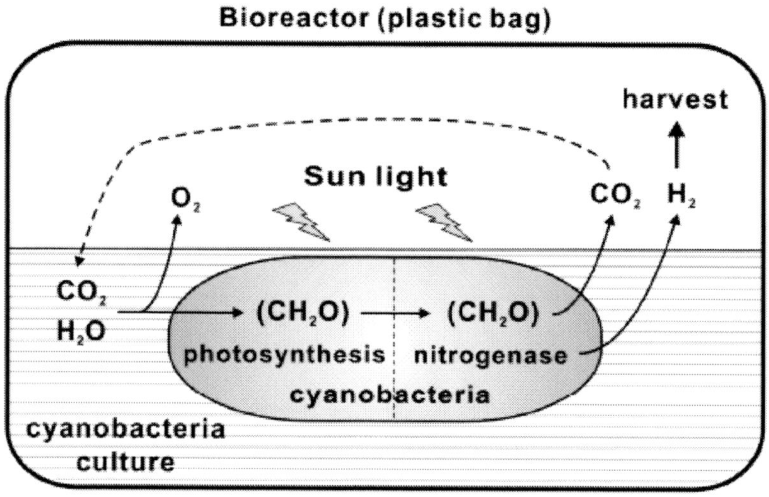

a)

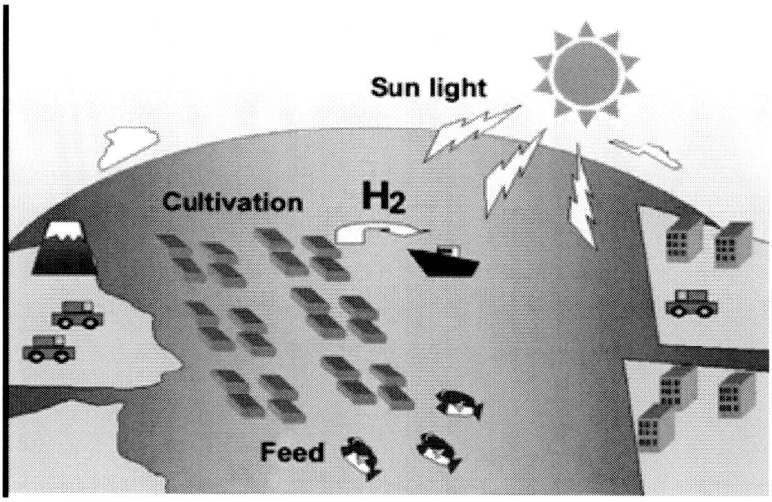

b)

Figure 2. Image of large-scale H_2 production utilizing mariculture-raised cyanobacteria. **A**. Bioreactor. The bioreactor (25 m wide and 200 m long) consists of 3 layers of plastic films: the inner-most bag holds the culture medium, cyanobacteria and gases, the middle one is a barrier to H_2, and the third one protects the inner ones from damage. For economical H_2 production, bioreactors should be inexpensive. **B**. Illustration of photobiological H_2 production on the sea surface. Vast sea surface areas are required in order to replace the energy currently supplied from fossil fuels and thus result in a meaningful emission reduction of CO_2.

Process B. Repeated harvesting of gas

When H_2 gas attains a certain concentration level (e.g. 25–30%, v/v), a factory ship harvests the gas mixture (main components: Ar, H_2, O_2) from bioreactors with hoses, and purifies H_2 on board. A group of smaller ships assist in the connection and disconnection of the hoses to the bioreactors (Figure 2B)

Process C. Separation and storage

The purification steps may consist of separation of H_2 using gas selective permeability membranes, followed by pressure-swing adsorption (PSA). The purified H_2 can be compressed, liquefied, or even transformed to hydrocarbons or alcohols with coal or CO_2 as the hydrogen acceptor. Ar gas separated from the combined gas is returned to bioreactors in Process A. A part of the energy required for these steps can be supplied by H_2.

Process D. Utilization

H_2 can be used as an energy source for fuel cells for electricity generation, vehicles (fuel cells or direct combustion), etc.

3. Basis of Biological Process

3.1. Choice of O_2 Evolving Photosynthetic Organisms

Many photosynthetic bacteria (e.g. purple bacteria, green sulfur bacteria) can convert solar energy into H_2. They cannot use H_2O as the electron donor and require organic compounds or sulfur compounds as the source of electrons for H_2 evolution (Rao and Cammack, 2001, Das and Veziroglu 2001), and the amount of the resource (electron donors) is limited. For large-scale photobiological H_2 production, the candidate microorganisms must use H_2O as the electron donor, thus narrowing the possibilities to cyanobacteria and eukaryotic microalgae (Rao and Cammack 2001; Prince and Kheshgi, 2005; Kruse et al., 2005; Sakurai and Masukawa, 2007).

3.2. Research Efforts Involving Photobiological H_2 Production by Using Naturally Selected Oxygen-Evolving Organisms

At the beginning of the oil crisis in the 1970s, tapping into alternative bio-solar energy production systems, including H_2 production was widely considered (Gibbs et al., 1973, Lien and San Pietro, 1975), but the enthusiasm of scientific community lasted for a relatively short period around 1975. During the two to three decades that followed the oil crisis, while our society paid little attention to the development of bio-solar energy conversion, some researchers still eagerly pursued the possibility of photobiological H_2 production. For example, Hallenbeck et al. (1978) found that some outdoor cultures of cyanobacteria were able to produce H_2 for more than 25 days under a constant stream of Ar-1% N_2-0.3% CO_2. Mitsui et al. (1983) collected a large number of photosynthetic microorganisms from natural environments (mostly from the Caribbean Sea), and found that some cyanobacteria (e.g. *Oscillatoria* sp. strain Miami BG7) were able to produce H_2 under continuous illumination.

The advances in research up to the middle 1990s in cyanobacteria as solar energy convertors were summarized (Hall et al. 1995). In order to improve H_2 productivity, a coiled tubular bioreactor was constructed (Tsygankov et al. 2002). When cyanobacteria use H_2O as the terminal electron donor for H_2 production, O_2 is concomitantly produced and may inhibit H_2 accumulation by reoxidizing H_2 through the activities of hydrogenase and the respiratory chain terminal oxidase. In order to prevent reoxidation of H_2 by O_2, a continuous flow of inert gas (lift gas) was employed in many of the above experiments. Some cyanobacterial strains obtained from natural habitats (Mitsui et al., 1983; Kumazawa and Mitsui, 1994; Kumazawa and Asakawa, 1995) and some generated by random chemical mutagenesis (Mikheeva et al., 1995) seem to be largely deficient in H_2 uptake activity, and were found to be suitable for photobiological H_2 production. These findings were important for demonstrating the potential for photobiological H_2 production using oxygen-evolving photosynthetic microorganisms along with some technological refinements.

3.3. The Enzymes of H_2 Evolution

The enzymes that can be used for H_2 production are either hydrogenase or nitrogenase (Review: Rao and Cammack, 2001; for cyanobacteia: Tamagnini et al. 2002). Hydrogenases that catalyze virtually unidirectional H_2 uptake are called uptake hydrogenases (Hup), and are not suitable for H_2 production. Hydrogenases that catalyze both production and uptake of H_2 are called reversible or bidirectional hydrogenases and can be used for H_2 production. Some species of eukaryotic microalgae and cyanobacteria have reversible hydrogenases and the reaction partners of these hydrogenases are ferredoxin, flavodoxin, and/or $NAD(P)^+/NAD(P)H$:

$$2\ H^+ + 2\ e^- \text{ (reduced ferredoxin, reduced flavodoxin, or NAD(P)H)} \leftrightarrow H_2 \text{ (+ oxidized donors).}$$

This type of hydrogenase in cyanobacteria is usually referred to as a bidirectional hydrogenase (Hox).

Nitrogenase is an enzyme of nitrogen fixation (Rees and Howard 2000, Igarashi and Seefeldt 2003) that typically catalyzes the following reaction with reduced ferredoxin/flavodoxin as electron donors:

$$N_2 + 8e^- + 8\ H^+ + 16\ ATP \rightarrow H_2 + 2\ NH_3 + 16\ (ADP + Pi) \qquad (1)$$

Nitrogenase occurs in prokaryotes only, and among oxygen-evolving photosynthetic organisms only some cyanobacteria (including prochlorophytes) have it.

3.4. Various Strategies for the Exploitation of Photobiological H_2 Production

3.4.1. Hydrogenase-Based System

Melis et al. (2000) found that the green alga *Chlamydomonas* cells evolved H_2 for a few days when photosystem II (PSII) that normally splits H_2O and evolves O_2, was apparently inactivated by culturing cells under sulfur-limited conditions. The observation was later explained as: 1) accumulation of organic substances by ordinary O_2-evolving photosynthesis, 2) sulfur deprivation leads to lowered PSII activity, 3) under microoxic conditions, e- supplied by H_2O and stored organic substances are used for H_2 production via hydrogenase, while O_2 produced is removed through respiration by consuming previously stored organic substances (Ghirardi et al. 2000, 2005, Happe et al. 2002, Melis 2002, Melis and Happe 2004).

3.4.2. Two-Stage Photo-H_2 Production

Benemann's group (Benemann 2000, Hallenbeck and Benemann 2002) proposes indirect processes (photofermentation) that consist of 1) production in open ponds of a biomass high in storage carbohydrates by ordinary photosynthesis, 2) concentration of the biomass, 3) anaerobic dark fermentation converting stored carbohydrate to H_2 and acetates, followed by 4a) photofermentation converting the by-product acetates to H_2 in closed photobioreactors, or 4b) high yields of H_2 via dark fermentation with carbohydrates generated through limited respiration. The latter process eliminates the need for anaerobic photobioreactors covering large areas.

3.4.3. Nitrogenase-Based System

The advantage of this reaction [Equation (1)] is that it irreversibly produces H_2 as an inevitable by-product even in the presence of O_2, if hydrogenase activity is absent. Thus, cyanobactria are able to continuously produce and accumulate H_2 by the collaboration between photosynthesis and nitrogenase, with H_2O as the ultimate electron donor accompanied by production of O_2. Its downside is that the reaction consumes a lot of the high-energy compound ATP, thus the theoretical maximum energy conversion is lower than the hydrogenase-based system. Under conditions ideal for nitrogen fixation (Equation (1)), 75% of electrons are allocated to nitrogen fixation and only 25% are allocated to H_2 production. However, the allocation coefficient can be modified in favor of H_2 production, and in the absence of N_2, all the electrons are allocated to H_2 production (see, Sakurai and Masukawa, 2007).

3.5. Development of Research for Nitrogenase-Based H_2 Production

3.5.1. More Detailed Description of Nitrogenase

Nitrogenases typically possess MoFeS clusters (Mo-type) as the catalytic centers (Einsle et al. 2002), but some bind V (V-type) or Fe (Fe-only type) instead of Mo (Eady 1996, 2003). The latter two types of nitrogenases are assumed to be less efficient in N_2 fixation as they allocate a higher ratio of electrons to H^+ reduction. In other words, they are assumed to be more suitable for further development in H_2 production systems. Some cyanobacteria have the V-type (Thiel, 1993) in addition to Mo-type, but the occurrence of the Fe-only type in cyanobacteria has yet to be proved. In the absence of N_2 (e.g. under Ar), all the electrons are allocated to H_2 production irrespective of the types of nitrogenase.

$$2\ e^- + 2\ H^+ + 4\ ATP \rightarrow H_2 + 4\ (ADP + Pi) \qquad (2)$$

Because nitrogenase is extremely O_2 sensitive, photolysis of H_2O and the direct coupling of the N_2ase reaction to reduced ferredoxin/flavodoxin are incompatible reactions. Cyanobacteria have evolved various strategies to spatially and/or temporally separate N_2ase from O_2 evolution activities, and these strategies have been used to classify the cyanobacteria into several groups (Gallon 2001, Tamagnini et al. 2002, Berman-Frank et al. 2003, Madamwar et al. 2000). Group A: filamentous heterocystous (e.g. *Nostoc*, *Anabaena*), B: filamentous non-heterocystous (*Oscillatoria, Trichodesmium*), and C: unicellular (*Cyanothece, Gloeothece*). Because the above antinomic situation is physiologically resolved by the cyanobacteria, frequent gas changes to sustain H_2 production are not necessarily needed, unlike the H_2ase-based production systems.

3.5.2. Improvement of Cyanobacteria by Genetic Engineering

We have been studying the nitrogenase-based H_2 production system of heterocyst forming cyanobacteria, because many of them are amenable to further refinement through genetic engineering. Nitrogen-fixing cyanobacteria may have up to two types of hydrogenase: the upake hydorogenase called Hup and the bidirectional hydrogenase called Hox. The presence of hydrogenases is considered to be one of the major obstacles to achieving efficient solar energy conversion in nitrogenase-based systems simply because the hydrogenases reabsorb the H_2 produced (Figure 3)

The framework for improvement of cyanobacteria using geneticc engineering was established by the late 1980s (Elhai and Wolk, 1988), and it has been a very effective strategy for improving H_2 productivity.

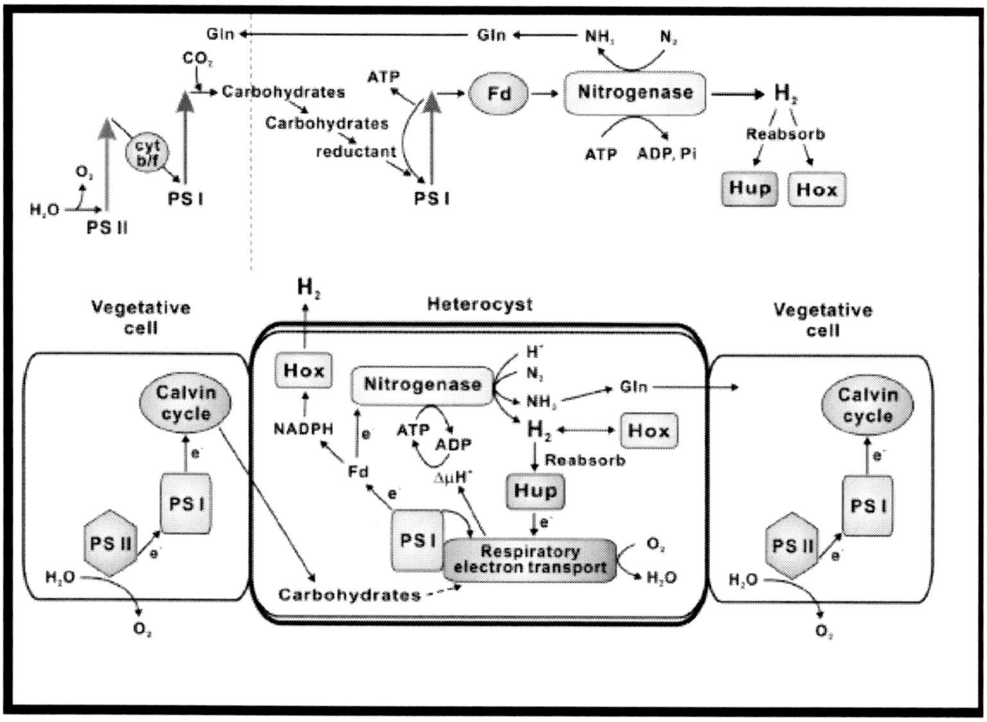

Figure 3. Energy metabolism of nitrogen fixing cyanobacteria (heterocyst forming type).

Because nitrogenase is sensitive to O_2 and therefore incompatible with O_2 evolving photosynthesis, the two activities are spatially separated in heterocyst-forming cyanobacteria. Vegetative cells synthesize carbohydrates from CO_2 and H_2O by ordinary photosynthesis (Calvin cycle), and send the carbohydrates to heterocysts specialized for nitrogen fixation. Heterocysts cannot perform O_2 evolution, and the electrons required for nitrogenase reaction are furnished by decomposition of the carbohydrates received from vegetative cells. H_2 is produced as an inevitable by-product in the nitrogenase reaction. Because the presence of hydrogenase activity reabsorbs H_2, it must be inactivated for efficient H_2 production. Fd: ferredoxin, Hox: bidirectional hydrogenase, Hup: uptake hydrogenase, PSI: photosystem I, PSII: photosystem II. (For details, see Sakurai and Masukawa, 2007.)

The work was further advanced with the publication of the entire genomic sequence of a nitrogen-fixing cyanobacterium in 2001 (Kaneko et al. 2001). Additionally, work in several laboratories demonstrated the feasibility of achieving enhanced H_2 production through genetic modifications. For example, Happe et al. (2000) created a genetically defined *hup* disruption mutant of *Anabaena variabilis* ATCC 29413, which produced H_2 at 4–6 times the rate of wild-type. In other work, Masukawa et al. (2002) constructed three genetically defined hydrogenase mutants including *hox* disrupted mutants (Δ*hupL*, Δ*hoxH*, and Δ*hupL*/Δ*hoxH*) from *Anabaena/Nostoc* sp. PCC 7120. Of these three mutants, two mutants, Δ*hupL* and Δ*hupL/ΔhoxH* produced H_2 at a rate 4–7 times that of wild-type under optimal conditions, while the effects of Hox inactivation were not evident. In similar work of Lindberg et al. 2002, disruption of the *hup* of in a *Nostoc punctiforme* ATCC 29133 strain that lacks Hox, was shown to be effective in enhancing H_2 production.

3.5.3. Recent Advancements in Our Laboratory

In order to establish strains with high levels of H_2 production, we have identified the promising wild-type strain *Nostoc* sp. PCC 7422 with high nitrogenase activity from among 13 nitrogen-fixing *Nostoc* and *Anabaena* strains. We subsequently improved H_2 producing activity of the cells by inactivating the uptake hydrogenase gene (*hupL*) via genetic engineering to create a Δ*hupL* mutant (Yoshino et al. 2007). The Δ*hupL* mutant cells were able to accumulate H_2 to about 22–29% even in the presence of evolved O_2.

More recently, we have found in a preliminary investigation that by improving the culture conditions we are able to harvest about 15% H_2 at least three times at an interval of 4 days each. We believe that there is the potential for continuous H_2 harvesting in such a system as the culture showed no appreciable signs of lowering its production activity by the conclusion of our experiment (unpublished observation).

3.5.4. Needs for further Improvement

The theoretically maximum energy conversion efficiency of H_2 production with H_2O as the electron donor is calculated to be 13.9–16.3% versus ordinary photosynthesis under visible light (Sakurai and Masukawa, 2007). On the other hand, the efficiency of ordinary photosynthesis is similarly calculated to be 27.6% for C3 plants such as rice and wheat, and 20.7–24.5% for C4 plants such as sugar cane and maize.

As visible light accounts for about 45% of total solar energy received on the Earth's surface, the above efficiency of H_2 production corresponds to 6.3–7.3% vs. total solar energy (Sakurai and Masukawa, 2007).

Under laboratory conditions, cyanobacteria sometimes produced H_2 with a light energy conversion efficiency of higher than 2.5% (versus visible light energy) (Kumazawa and Mitsui, 1994; Yoshino et al., 2007). However, these high efficiencies are only attained in a relatively short period (several days) under low light intensities of about one twenty-fifth of full sunlight received on the equator. Under outdoor conditions, the reported best value of the efficiency in a relatively long period (days) is about 0.1% (Tsygankov et al., 2002). More research is needed to improve the outdoor light energy conversion efficiency (for a review, see Sakurai and Masukawa, 2007)

We believe that the potential exists for improving the outdoor efficiency of H_2 production for the following reasons:

1) In shorter-term (illumination period, 3 or 4 hours) experiments, several green algae produced H_2 (by hydrogenase) at an energy conversion efficiency (ΔG) of 6–24 % (about 7–28% in ΔH) vs. visible light. (Greenbaum, 1988).
2) The green alga *Chlorococcum lottorale* grew photosynthetically (C3 type) in a vertical flat-plate photobioreactor outdoors at a light energy conversion efficiency of about 5.9–6.3% (vs. visible light) in a series of single-day experiments that took into account night respiration loss (bubbling with air containing 5% CO_2, irradiation (visible light): 8.3–11.2 MJ m^{-2} day^{-2} or 2.3-3.1 kWh m^{-2} day^{-2} in Winter, Spring, and Summer, in Japan) (Zhang et al. 2001).
3) From the ratio between photobiological H_2 production and C3 type photosynthesis of the theoretical maximum efficiency (13.9–16.3% vs. 27.6%), the efficiency of 5.9–6.3% (average 6.2%) for the latter corresponds to 3.1–3.7% versus visible light or 1.4–1.7% versus total solar radiation for the former (H_2 production).

3.5.5. Our Tentative Energy Conversion Efficiency Target for H_2 Production Systems

In order to demonstrate the potential of large-scale photobiological H_2 production using cyanobacteria, we set our tentative target of solar energy conversion efficiency at 0.5% versus total solar irradiance outdoors, and from the above considerations, we think that this is an attainable goal. Achieving efficiencies of 1% and higher will be the next goal. In order to further improve efficiencies, one of the most important considerations is the inherent low-light saturation property of most cyanobacteria strains that leads to low energy conversion efficiency outdoors. The strategy of selecting wild-type strains with high nitrogenase activity under full sunlight, followed by genetic engineering, should also be attempted as a means to overcome this limitation (Sakurai and Masukawa, 2007).

4. Trial Cost Analysis of H_2 Production

In the following calculations the cost of photobiological H_2 production is estimated based on somewhat subjective assumptions as a guide to readers in considering the feasibility of large-scale H_2 production, at present or in the future. We believe that in the future as new H_2

production, purification and separation technologies are developed, further refinement of the calculations is possible by adding new cost items and changing some of the assumed values.

4.1. Cost of Biological Production Stage

Cost of the Culture Medium

Many of the cyanobacteria strains can grow in liquid media that is relatively simple in composition. Cyanobacteria are cultured in liquid medium 20 cm in depth (200 l m^{-2} or 0.2 ton (metric ton) m^{-2}) utilizing either natural fresh water or seawater depending on the strain used. Potentially growth-limiting nutritional elements are added to the medium as 'fertilizers' akin to agricultural practices. An example of the composition of a synthetic medium (Rippka et al. 1979) frequently used in cyanobacteria culture is shown in Table 2. In our proposed system, $NaNO_3$ can be omitted because we use nitrogen-fixing cyanobacteria. If 0.18 mM K_2HPO_4 and 0.03 mM $FeCl_3$ (instead of ferric ammonium citrate) are added to the natural water, the cost of chemicals is about 5–20 cents (in U.S.A. currency) per 0.2 kl or the sunny side surface area of 1 m^2 of the bioreactor [Cost 1A]. Once grown, cyanobacteria continuously produce H_2 allowing repeated harvesting, further additions of nutrients and CO_2 are not necessary.

The Cost of Water

The water for the medium should be sufficiently purified by removing suspended particles including bacteria from natural water. The cost will vary by the availability and quality of water, and is difficult to estimate. As a reference price, water for industrial use are sold by local governments in Japan at about 7.5–60 cents per ton, which corresponds to 1.5–12 cents m^{-2} of the bioreactor surface (20 cm in depth leads to 0.2 kl) [Cost 1B].

The Cost of Bioreactor

As the substrate for H_2 production, 18 g of H_2O can generate 22.4 l (STP) of H_2, which corresponds to 1 kg of H_2O being converted to 1.24 m^3 of H_2, with an energy content (*H*: enthalpy) equivalent to about 0.4 l of crude oil (3 m^3 H_2 is equivalent to about 1 l of crude oil in enthalpy). The cost of substrate water used as the electron donor is thus negligible.

The bioreactor is composed of three layers of plastic bags, a total of 6 layers (sunny side and shady side) of plastic film altogether. The innermost bag holds cyanobacteria culture, the middle bag has very low permeability to H_2, and the outermost bag serves to mechanically protect the inner bags.

Cost of Gas

The bioreactor is initially filled with 94% Ar, 5% CO_2, and 1% N_2 at a total volume of 0.5 m^3 per m^2 of the reactor surface (initial gas phase: 0.5 m thick). When H_2 concentration reaches 30% (v/v) (0.28 l at STP, or 3.5 MJ in enthalpy equivalent to 0.09 l of crude oil), the total gas is harvested for H_2 separation. The frequency of the harvesting is once in about 25 days ($3.5/(14.5 \times 10^{-2}) = 24$). The price (FOB, bulk rate) of Ar is assumed to be \$100–500 per ton (cost of 9 cents – 45 cents per m^2 [Cost 1D]). The Ar obtained during separation of H_2 can be returned to the bioreactor. The quality of Ar gas needs not be high, because contaminating O_2

is consumed by respiration, and contaminating N_2 is consumed by nitrogen fixation. The price of CO_2 is assumed to be $4 – 15 per ton, therefore the total cost is negligible.

Table 2. Composition of the medium (BG11) frequently used for cyanobacteria culture (Rippka et al., 1979)

1) Macro-nutrients	
$NaNO_3$	17.65 mM
K_2HPO_4	0.18 mM
$MgSO_4$	0.3 mM
$CaCl_2$	0.25 mM
Na_2CO_3	0.19 mM
Ferric ammonium citrate	0.03 mM
Na_2,Mg-EDTA	0.003 mM
Citric acid	0.029 mM
2) Micronutrients	
B (46 μM), Mn (9.2 μM), and Co, Cu, Mo, Zn (each lower than 2 μM)	

Assumptions for cost estimation: 1) the thickness of each film is 0.08 mm, 480 cm^3 of plastic per m^3 of the bioreactor's sunny side surface is required as the feed-stock, 2) assuming the average plastic price of $2 – $4 per kg, results in about $1–$2 per m^3 of bioreactor, 3) the used plastics can be recycled many times to generate new plastic films at about half the price of the new materials. The above assumptions result in the cost of the bioreactor to be about 50 cents – $1 per m^3 of bioreactor surface [Cost 1C].

Other Cost of Biological Production Stage

In addition to the Costs A-D, the following costs will be incurred in the biological H_2 production stage: the cost of growing cyanobacteria, the labor cost, the cost of ships, the interest on capital goods, and the cost of marine transportation of production materials. [Cost 1E].

The energy required for the biological H_2 production stage is assumed to be covered by the H_2 produced [Energy Cost 1A] (see below).

4.2. Cost of Gas Harvesting, H_2 Separation, Storage, and Marine Transportation to Final Destination.

As many of the relevant technologies are under development, here we simply assume that one third [Case A] and a half [Case B] of energy produced in the biological production stage is consumed in these processes [Energy Cost 2A] plus [Energy Cost 1F].

As in the first stage, the labor costs, the cost of capital goods and the interest on them, and transportation costs are also required [Cost 2B].

4.3. Total Cost

1) Final yield of H_2
2) At 1% light conversion efficiency, H_2 energy of 15 kWh (or 54 MJ) m^{-2} $year^{-1}$ is produced, and 36 ([Case A]) and 27 ([Case B]) MJ m^{-2} $year^{-1}$, respectively, are obtained as the final product H_2.

3) Cost 1A-D
4) The sum of Cost 1A-D is 65–180 cents m^{-2} $year^{-1}$. Assuming the above final yield 1), the cost of H_2 is 1.8–7.8 ([Case A]) and 2.4–10.4 cents MJ^{-1} ([Case B]).
5) Cost 1E plus Cost 2A
6) The values of these costs are left blank, because we have no reliable estimates at this moment. Note that we need not include the energy cost here, because it is considered in 1).

Final calculation

I. I. If we assume that Cost 1E plus Cost 2A are a half of Cost 1A-D, the final price of H_2 is 2.7-7.5 ([Case A]) and 3.6-10.0 cents MJ^{-1} ([Case B]).

II. If we assume that Cost 1E plus Cost 2A are equal to Cost 1A-D, the final price of H_2 is 3.6-10.0 ([Case A]) and 4.8-13.4 cents MJ^{-1} ([Case B]).

These values are compared with the cost ranges of other renewable energy sources in Table 3.

Table 3. Comparison of price of renewable energy sources

	Current cost ($ct MJ-1)	Potential long-term Future cost (2050) ($ct MJ-1)	Theoretical potential (EJ)
Photo-bio H_2			27,000 [1)]
Case A-I		1.8-5.0	
Case A-II		3.6-10.0	
Case B-I		2.4- 6.7	
Case B-II		4.8-13.4	
Wind electricity [2)]	2.0-5.2	1.2-4.0	6,000
Solar Photovoltanic electricity [2)]	10-50	2-10	540,000 [3)]
(crude oil of $50-150 per barrel)	(0.8-2.4)		

(assumption: 1 Euro = 1.45 US dollar)
[1)] at 1% energy conversion efficiency, [2)] IEA (2007a), [3)] at 20% energy conversion efficiency

Conclusion

We briefly described the future prospects of our proposal for large-scale H_2 production by mariculture. We believe that this proposal is worth pursuing as both the energy source (Sun) and the feed stock (water) are almost limitless, and the use of these resources does not compete with food production such as land biomass. In addition, in contrast with electricity generated by photovoltanics or wind, H_2 is easily stored and transported overseas. Our trial calculation indicates that bio-solar H_2 production will not be competitive with fossil fuels at current prices, but as specific policies come into play such as carbon taxes, there will likely be an acceleration of its development. According to the IPCC report: “Depending on the existing tax system and spending of the revenues, modeling studies indicate that costs (for mitigation of climate change) may be substantially lower under the assumption that revenues from carbon taxes or auctioned permits under an emission trading system are used to promote low-carbon technologies or reform of existing taxes.", and "Policies that provide a real or implicit price of carbon could create incentives for producers and consumers to significantly invest in low-GHG products, technologies and processes. Such policies could include economic instruments, government funding and regulation". (IPCC, 2007c)

In spite of the technical as well as economical difficulties, we propose that large-scale H_2 production by mariculture is the most promising strategy to be pursued for mitigating dangerous anthropogenic interference with the climate system that currently threatens present and future generations with the continuation of the rising atmospheric CO_2 levels.

Acknoweledgement

We thank Dr. Susan Carlson for carefully reading the manuscript. This work was supported in part by the MEXT grant Venture (2001-2005) to HS, Grants for Special Research Projects, Waseda University (2004A-085, 2005B-083) to HS, Grants-in-Aid for Scientific Research from JSPS to HM (16-9494) and MEXT Grant-in-aid for the High-Tech Research Center Project (2007) to KI

References

Berman-Frank, I., Lundgren, P. and Falkowski, P. (2003) Nitrogen fixation and photosynthetic oxygen evolution in cyanobacteria. Res. Microbiol. 154, 157-164.

Benemann, J.R. (2000) Hydrogen production by microalgae. *J. Appl. Phycol.* 12, 291-300.

Das, D. and Veziroglu, T.N. (2001) Hydrogen production by biological processes: a survey of literature. *Int. J. Hydrogen Energy* 26, 13-28.

Eady, R.R. (1996) Structure-function relationships of alternative nitrogenases. *Chem. Rev.* 96, 3013-3030.

Eady, R.R. (2003) Current status of structure function relationships of vanadium nitrogenase. *Coordination Chem. Rev.* 237, 23-30.

Einsle, O., Tezcan, F.A., Andrade, S.L.A., Schmid, B., Yoshida, M., Howard, J.B. and Rees, D.C. (2002) Nitrogenase MoFe-protein at 1.16 angstrom resolution: A central ligand in the FeMo-cofactor. *Science* 297, 1696-1700.

Elhai, J. and Wolk, C.P. (1988) Conjugal transfer of DNA to cyanobacteria. *Methods Enzymol.* 167, 747-754.

Gallon, J.R. (2001) N_2 fixation in phototrophs: adaptation to a specialized way of life. *Plant Soil* 230, 39-48.

Ghirardi, M.L., Zhang, J.P., Lee, J.W., Flynn, T., Seibert, M., Greenbaum, E. and Melis, A. (2000) Microalgae: a green source of renewable H_2. *Trends Biotechnol.* 18, 506-511.

Ghirardi, M.L., King, P.W., Posewitz, M.C., Maness, P.C., Fedorov, A., Kim, K., Cohen, J., Schulten, K. and Seibert, M. (2005) Approaches to developing biological H_2-photoproducing organisms and processes. Biochem. *Soc. Transactions* 33, 70-72.

Gibbs, M., Hollaender, A., Kok, B., Krampitz, L.O. and San Pietro, A., Proceedings of the Workshop on Bio-Solar Energy Conversion (1973, September) NSF Report (RANN Grant GI 40253), NSF, Bethesda Md., 1973.

Greenbaum, E. (1988) Energetic efficiency of hydrogen photoevolution by algal water splitting. *Biophys. J.* 54, 365-368.

Hall, D.O., Markov, S.A., Watanabe, Y. and Rao, K. (1995) The potential applications of cyanobacterial photosynthesis for clean technologies. *Photosyn. Res.* 46, 159-167.

Hallenbeck, P.C. and Benemann, J.R. (2002) Biological hydrogen production; fundamentals and limiting processes. *Int. J. Hydrogen Energy* 27, 1185-1193.

Hallenbeck, P.C., Kochian, L.V,. Weissman, J.C. and Benemann, J.R. (1978) Solar energy conversion with hydrogen-producing cultures of the blue-green alga, *Anabaena cylindrica*. Biotechnol. *Bioengineer. Symp.* No 8, 283-297.

Happe, T., Hemschemeier, A., Winkler, M. and Kaminski, A. (2002) Hydrogenases in green algae: do they save the algae's life and solve our energy problems? *Trends Plant* Sci. 7, 246-250.

Happe, T., Schütz, K. and Böhme, H. (2000) Transcriptional and mutational analysis of the uptake hydrogenase of the filamentous cyanobacterium *Anabaena variabilis* ATCC 29413. *J. Bacteriol.* 182, 1624-1631.

IEA, (2007a) Bioenergy, Potential Contribution of Bioenergy to the World's Future Energy Demand [cited 2007, Nov. 20th]. Availabl from: URL: http://www.ieabioenergy.com/

IEA, (2007b) World Energy Outlook 2007 [cited 2007, Nov. 20th]. Availabl from: URL: http://www.worldenergyoutlook.org/

Igarashi, R.Y. and Seefeldt, L.C. (2003) Nitrogen fixation: The mechanism of the Mo-dependent nitrogenase. *Crit, Rev, Biochem, Mol, Biol*, 38, 351-384.

IPCC, (2007a) Press Release, Summary of Speech by Dr. R K Pachauri, Chairman, at the Opening of The Ninth Session of IPCC Working Group III at Bangkok, April 29, 2007. [cited 2007, Nov. 20th]. Availabl from: URL: http://www.ipcc.ch/pdf/press-ar4/wg3/speech-29april2007-rajendra-pachauri.pdf

IPCC, (2007b) The 4th Assessment Report, Working Group I. Summary for Policy Makers. (2007) [cited 2007, Nov. 20th]. Availabl from: URL:
http://www.ipcc.ch/pdf/assessment-report/ar4/wg1/ar4-wg1-smp.pdf

IPCC, (2007c) The 4th Assessment Report, Working Group III. Summary for Policy Makers. (2007) [cited 2007, Nov. 20th]. Availabl from: URL:http://www.ipcc.ch/pdf/assessment-report/ar4/wg3/ar4-wg3-smp.pdf

Kaneko, T., Nakamura, Y., Wolk, C.P., Kuritz, T., Sasamoto, S., Watanabe, A., Iriguchi, M., Ishikawa, A., Kawashima, K., Kimura, T., Kishida, Y., Kohara, M., Matsumoto, M., Matsuno, A., Muraki, A., Nakazaki, N., Shimpo, S., Sugimoto, M., Takazawa, M., Yamada, M., Yasuda, M., Tabata, S. (2001) Coplete genomic sequence of the filamentous nitrogen-fixing cyanobacterium Anabaena sp. Strain PCC 7120. *DNA Res.* 8, 227-253

Kheshgi, H.S., Prince, R.C. and Marland, G. (2000) The potential of biomass fuels in the context of global climate change: Focus on transportation fuels. *Ann. Rev. Energy Environ.* 25, 199-244.

Kruse, O., Rupprecht, J., Mussgnug, J.R., Dismukes, G.C. and Hankamer, B. (2005) Photosynthesis: a blueprint for solar energy capture and biohydrogen production technologies. *Photochem. Photobiol.* Sci. 4, 957-970.

Kubota, M., Takayama, K. and Namimoto, D. (2005) Pleading for the use of biodegradable polymers in favor of marine environments and to avoid an asbestos-like problem for the future. *Appl. Microbiol. Biotechnol.* 67, 469-476.

Kumazawa, S. and Mitsui, A. (1994) Efficient hydrogen photoproduction by synchronously grown cells of a marine cyanobacterium, *Synechococcus* sp. Miami BG-043511, under high cell-density conditions. *Biotechnol. Bioengineer.* 44, 854-858.

Kumazawa, S. and Asakawa, H. (1995) Simultaneous production of H_2 and O_2 in closed vessels by marine cyanobacterium *Anabaena* sp. TU37-1 under high-cell-density conditions. *Biotechnol. Bioengineer.* 46, 396-398.

Levin, D.B., Pitt, L. and Love, M. (2004) Biohydrogen production: prospects and limitations to practical application. *Int. J. Hydrogen Energy* 29, 173-185.

Lien, S. and San Pietro, A., An Inquiry into Biophotolysis of Water to Produce Hydrogen, NSF Report (RANN Grant GI 40253), NSF, Bethesda Md., (publication date, not described. Deemed to be around) 1975.

Lindberg, P., Schütz, K., Happe, T. and Lindblad, P. (2002) A hydrogen-producing, hydrogenase-free mutant strain of *Nostoc punctiforme* ATCC 29133. *Int. J. Hydrogen Energy* 27, 1291-1296.

Madamwar, D., Garg, N. and Shah, V. (2000) Cyanobacterial hydrogen production. *World J. Microbiol. Biotechnol.* 16, 757-767.

Masukawa, H., Mochimaru, M. and Sakurai, H. (2002) Disruption of the uptake hydrogenase gene, but not of the bidirectional hydrogenase gene, leads to enhanced photobiological hydrogen production by nitrogen-fixing cyanobacterium *Anabaena* sp. PCC 7120. *Appl. Microbiol. Biotechnol.* 58, 618-624.

Melis, A. (2002) Green alga hydrogen production: progress, challenges and prospects. *Int. J. Hydrogen Energy* 27, 1217-1228.

Melis, A. and Happe, T. (2004) Trails of green alga hydrogen research - from Hans Gaffron to new frontiers. *Photosyn. Res.* 80, 401-409.

Melis, A., Zhang, L.P., Forestier, M., Ghirardi, M.L. and Seibert, M. (2000) Sustained photobiological hydrogen gas production upon reversible inactivation of oxygen evolution in the green alga *Chlamydomonas reinhardtii*. *Plant Physiol.* 122, 127-135.

Mertens, R. and Liese, A. (2004) Biotechnological applications of hydrogenases. *Curr. Opinion Biotechnol.* 15, 343-348.

Mikheeva, L.E., Schmitz, O., Shestakov, S.V. and Bothe, H. (1995) Mutants of the cyanobacterium *Anabaena variabilis* altered in hydrogenase activities. *Z. Naturforsch.* 50c, 505-510.

Mitsui, A., Phlips, E.J., Kumazawa, S., Reddy, K.I., Ramachandran, S., Matsunaga, T., Haynes, L. and Ikemoto, H. (1983) Progress in research toward outdoor biological hydrogen-production using solar-energy, sea-water, and marine photosynthetic microorganisms. *Ann. N. Y. Acad. Sci.* 413, 514-530.

Murray, R.L. and Mitsui, A. (1982) Growth of hybrid tilapia frey fed nitrogen fixing marine blue-green algae in seawater. *J. World Maricul. Soc.* 13, 198-209.

Prince, R.C. and Kheshgi. H.S. (2005) The photobiological production of hydrogen: Potential efficiency and effectiveness as a renewable fuel. *Crit. Rev. Microbiol.* 31, 19-31.

Rao, K.K., and Cammack, R. (2001) Producing hydrogen as a fuel. In R. Cammack, M. Frey, and R. Robson (Eds.), *Hydrogen as a Fuel – Learning from Nature* (pp. 201-230). London and New York, Taylor and Francis.

Rees D.C. and Howard, J.B. (2000) Nitrogenase: standing at the crossroads. *Curr. Opinion Chem. Biol.* 4, 559-566.

Rippka, R. (1979) Genetic assignments, strain histories and properties of pure cultures of cyanobacteria. *J. Gen. Microbiol.* 111, 1-61.

Sakurai, H. and Masukawa, H. (2007) Promoting R and D in photobiological hydrogen Production utilizing mariculture-raised cyanobacteria. *Mar. Biotechnol.* 9, 128-145.

Tamagnini, P., Axelsson, R., Lindberg, P., Oxelfelt, F., Wünschiers, R. and Lindblad, P. (2002) Hydrogenases and hydrogen metabolism of cyanobacteria. *Microbiol. Mol. Biol. Rev.* 66, 1-20.

Tsygankov, A.A., Fedorov, A.S., Kosourov, S.N. and Rao, K.K. (2002) Hydrogen production by cyanobacteria in an automated outdoor photobioreactor under aerobic conditions. *Biotechnol. Bioengineer.* 80, 777-783.

Thiel, T. (1993) Characterization of genes for an alternative nitrogenase in the cyanobacterium *Anabaena variabilis. J. Bacteriol.* 175, 6276-6286.

UNFCCC (2007) home page, [cited 2007, Nov. 20th]. Availabl from: URL:http://unfccc.int/2860.php

Yoshino, F., Ikeda, H., Masukawa, H. and Sakurai. H. (2007) High photobiological hydrogen production activity of a *Nostoc* sp. PCC 7422 uptake hydrogenase-deficient mutant with high nitrogenase activity. *Mar. Biotechnol.* 9, 101-112.

Zhang, K., Miyachi, S. and Kurano, N. (2001) Photosynthetic performance of a cyanobacterium in a vertical flat-plate phtobioreactor for outdoor microalgal production and fixation of CO_2. *Biotechnol. Lett.* 23, 21-26.

In: New Research on Energy Economics
Editors: P.G. Caldwell and E.V. Taylor, pp. 169-182
ISBN: 978-1-60456-354-2

Chapter 6

DYNAMIC RELATIONSHIPS AMONG OIL PRICE SHOCK, STOCK MARKET RETURNS AND ECONOMIC ACTIVITY: THEORETICAL AND EMPIRICAL EVIDENCE

Evangelia Papapetrou[1*]
University of Athens, Department of Economics, Athens, Greece
and Bank of Greece, Economic Research Department

Abstract

Using a multivariate vector-autoregression (VAR) approach, the paper studies the dynamic relationship among oil prices, real economic activity, real stock prices and real interest rates in Greece over the period 1989:1 to 2007:6. The empirical evidence suggests that oil price changes affect real economic activity and are important in explaining stock price movements. Employing vector error-correction models and generalized impulse response functions, the empirical results support the endogeneity of real economic activity and real stock prices however the signal transmitted from one variable to the other is weak.

Keywords : Oil prices; stock returns; output.

Introduction

The correlation between oil prices and real output has received considerable attention especial for most industrialized countries (Hamilton, 1983; Burbridge and Harrison, 1984; Gisser and Goodwin, 1986; Mork, 1989; and Ferderer, 1996; Guo and Kliesen, 2005). One explanation for such a relationship is that oil prices increases lower future GDP growth by raising

[1] E-mail address: epapapetrou@bankofgreece.gr. Tel.: +301 320. 22377 Fax: +301 323.3025. Address for correspondence: Evangelia Papapetrou, Economic Research Department, Bank of Greece, 21 El. Venizelou St., 102 50 Athens, Greece.
[*] The views expressed in this paper are those of the author and not those of the Bank of Greece.

production costs. Various researchers have documented a negative relationship between price changes and economic activity. In a seminal work, Hamilton (1983) showed that oil price increases are responsible for almost every post World War II US recession. Later other researchers extended Hamilton's basic findings using alternative data and estimation procedures (Burbridge and Harrison, 1984; Gisser and Goodwin, 1986; Olson, 1988).[2]

Recently changes in the price of oil are considered as an important factor explaining fluctuations in stock prices. However, there is no consensus among researchers about the relation between oil prices and stock prices. Researchers, such as Chen, Roll and Ross (1986) show that oil prices have no effect on asset pricing. Jones and Kaul (1996) find that a negative relationship between stock prices and oil price changes exists for the U.S but they claim that the results for both the United Kingdom and Japan are not as strong. Huang, Masulis and Stoll (1996) found that oil future returns do not have much impact on general market indices. Kilian and Park 2007 conclude that the negative relationship of stock prices and oil price shocks is present when the oil price rises due to an oil-market specific demand shock. However, when higher oil prices are the outcome of a global economic expansion, then a persistent positive effect on cumulative stock returns is reported.

Nevertheless, the analysis of the relation between oil price changes, economic activity and stock prices outside the U.S has received little attention. Few studies have examined the effects of oil shocks on the stock market and economic activity mainly for a few industrialized countries such as the United Kingdom, Japan and Canada (Mork, Olsen and Mysen, 1994) and recently Greece (Papapetrou, 2001) and Spain (Miguel, Manzano and Martin-Moreno, 2003).

The purpose of the paper is to contribute to the empirical investigation of the oil price changes, economic activity and stock prices relationship. Utilizing time-series methods and applying vector error-correction (VEC) model estimation, the paper studies the dynamic interactions among oil prices, stock returns, interest rates and economic activity in Greece, over the period 1989:1 to 2007:6. Greece is a medium-sized economy heavily dependent on oil, and is more vulnerable than other economies to changes in the international oil market. Moreover Greece is a European Union country with a less mature financial market compared to other advanced economies. Over the last two decades, its financial market has undergone a process of liberalization at an accelerating pace and expanded considerably, while the fairly remarkable growth rates achieved by the Greek economy after the early nineties have enabled the country to enter the Euro zone.

The paper proceeds as follows. Section II presents a review of previous studies on the empirical evidence of oil price changes and its affects on stock returns and economic activity. Section III deals with methodological issues and the data used in the empirical analysis, while in Section IV the empirical evidence is presented. Finally, in Section V, the conclusions of the analysis are summarized.

Oil Prices, Output and Stock Prices: Evidence from the Literature

The negative correlation between oil prices and real output has been addressed by various researchers. Hamilton (1983) using Granger causality examined the impact of oil price shocks

2 For an extensive review on the oil price shocks and macroeconomy and a discussion on various transmission channels of oil price shocks, see among others, Jones et. al. (2004) and Cologni and Manera (2007).

and the US economy in 1949-1972 and 1973-1980. Hamilton finds that changes in oil prices Granger-caused changes in GNP whereas oil prices were determined exogenously.

Burbidge and Harrison (1984) examine the impact of oil price shocks on some macroeconomic variables in the U.S.A., Canada, U.K., Japan and Germany. Using VAR models they show that the 1973-74 oil embargo explains a substantial part of the behavior of industrial production in each of the countries examined. They reach the same conclusions as in Hamilton's work. However, for the oil changes in 1979-80 they find little evidence that the changes in oil prices have an effect on industrial production.

The existence of an asymmetric relationship between oil price changes and economic output was studied by various researchers. Mork (1989) extends Hamilton's analysis, by including the oil price collapse of 1986. He confirms Hamilton's (1983) results by finding that a strong, negative correlation between oil price increases and the growth of GNP for the United States persists when the sample is extended beyond the 1985-86 oil price decline. Moreover, the coefficients on oil price increases and oil price decreases were significantly different from each other indicating that the effects if oil price increases and decreases were asymmetric.

The study of Mork, Olsen and Mysen (1994) extents the findings of the previous studies to include six other industrialized countries, namely, Japan, Germany (West), France, Canada, the United Kingdom, and Norway. These countries vary considerably in the degree to which their economies depend on oil as an input and to which they are dependent on foreign oil. They conclude that the negative correlation between oil price increases and growth in GDP is present and significant for most of the countries examined for data extending through 1992. Moreover, for most countries they show evidence of asymmetric effects.

Hamilton (1996) proposed a transformation of raw oil prices, as an alternative specification for oil price changes. In particular, an oil shock is equal to the difference between the current oil price and the maximum price of the previous four or twelve quarters. Hamilton finds that this alternative oil shock measure shows a negative and stable relation with future GDP growth.

Guo and Kliesen (2005) employing a volatility measure constructed using daily crude oil futures prices show that it has a negative and significant effect on future GDP growth and other measures of the U.S. macroeconomy over the period 1984-2004. Their results are consistent with the nonlinear effect documented in Hamilton's work (1996). They also find that macroeconomic variables do not forecast realized oil price volatility.

However, changes in the price of crude oil are often considered an important factor for understanding fluctuations in stock prices. Various studies have examined the relationship between oil price shocks and financial markets (Chen, Roll and Ross, 1986; Jones and Kaul, 1996; Haug et.al., 1996; Sadorsky, 1999; Kilian and Park 2007).

Chen, Roll and Ross (1986) document that oil price changes have no effect on asset pricing. Jones and Kaul (1996) test whether the reaction of international stock markets to oil shocks can be justified by current and future changes in real cash flows and/or changes in expected returns. They find that in the postwar period, the reaction of United States and Canadian stock prices to oil shocks can be completely accounted for by the impact of these shocks on real cash flows. In contrast, the results for both the United Kingdom and Japan are not as strong.

Huang, Masulis and Stoll (1996) examine the link between daily oil future returns and daily United States stock returns. The evidence suggests that oil futures returns do lead some individual oil company stock returns but oil future returns do not have much impact on general market indices.

Sadorsky (1999) using monthly data over the period 1947-1996 estimates various VARs of industrial production, real oil prices, short-term interest rates and an index of real stock returns on the SandP 500. He shows that oil prices and oil price volatily both play important roles in affecting real stock returns. Especially, after 1986 oil price movements explain a larger fraction of the forecast error variance in real stock returns than do interest rates.

In a recent study, Kilian and Park (2007) show that the responses of real U.S. stock returns to oil price shocks differ substantially, depending on the underlying causes of the oil price increase. They find that a negative response of stock prices to oil price shocks is found only when the oil price rises due to an oil-market specific demand shock, such as an increase in precautionary demand driven by fears about future availability of crude oil, is the underlying driving force. On the contrary, if higher oil prices are the outcome of a global economic expansion, then they report persistent positive effects on cumulative stock returns.

However, fewer studies have examined the relationship for other countries except the U.S. (Papapetrou, 2001; Miguel, Manzano and Martin-Moreno, 2003). Papapetrou (2001) using a VAR approach among oil prices, real stock prices, interest rates, real economic activity and employment in Greece shows that oil price changes affect real economic activity and that they are important in explaining stock price movements. Miguel, Manzano and Martin-Moreno (2003) use a dynamic general equilibrium model for Spain. Their results suggest that oil shocks account for more than half the size of the aggregate fluctuations of the economy. They show that increases in the relative price of oil had a negative and significant effect on welfare.

Methodological Issues and Data

In the empirical analysis a VAR analysis is performed to explain oil price changes and its effects on stock prices and economic activity. This method of analysis allows testing the endogeneity of all variables in the economy and estimating the responses of stock pieces and economic activity to oil prices shocks in order to capture the long-run and short-run dynamics of the variables. The specification employed in the empirical analysis consists of four variables: oil prices, real interest rate and real stock variables. The specification employs the industrial production as a proxy to capture the variations in economic activity.

Testing for the existence of statistical relationship among the variables is done in three steps. The first step is to verify the order of integration of the variables since the causality tests are valid if the variables have the same order of integration. Standard tests for the presence of a unit root based on the work of Dickey and Fuller (1979, 1981), Perron (1988), Phillips (1987) and Phillips and Perron (1988)[3] and Kwiatkowski *et al.* (1992)[4] are used to investigate the degree of integration of the variables used in the empirical analysis.

3 This version of the test is an extension of the Dickey-Fuller test which makes a semi-parametric correction for autocorrelation and is more robust in the case of weakly autocorrelated and heteroskedastic regression residuals. According to Choi (1992) the Phillips-Perron tests (PP) extension appear to be more powerful than the ADF tests for aggregate data. For more details see Perron (1990).

4 The KPSS procedure assumes the univariate series can be decomposed into the sum of a deterministic trend, random walk and stationary I(0) disturbance and is based on a Lagrange Multiplier score testing principle. This test reverses the null and the alternative hypothesis. A finding favorable to a unit root in this case requires strong evidence against the null hypothesis of stationarity.

The second step involves testing for cointegration using the Johansen maximum likelihood approach (Johansen 1988, Johansen and Juselius 1990, 1992). The Johansen-Juselius estimation method is based on the error-correction representation of the VAR(p) model with Gaussian errors.

If a long-run relationship exists among the four endogenous variables, then a VEC system is estimated where the estimated long-run relationship is the error-correction term. Hence, the empirical analysis involves utilization of the vector error-correction modeling and testing for exogeneity of variables. Engle and Granger (1987) show that in the presence of cointegration, there always exist a corresponding error-correction representation which implies that changes in the dependent variable are a function of the level of disequilibrium in the cointegrating relationship, captured by the error-correction term (ECT), as well as changes in other explanatory variables. Through the ECT, the VECM modeling establishes an additional channel to examine the Granger-causality ignored initially from the Granger-Sims tests. The Wald-test applied to the joint significance of the sum of the lags of each explanatory variable and to the lagged error-correction term will imply statistically the Granger-exogeneity or endogeneity of the dependent variable. The non-significance of ECT is referred to as long-run non-causality, which implies that the variable is weakly exogenous with respect to long-run parameters. The absence of short-run causality (Granger-causality in the original sense) is established from the non-significance of the lags of each explanatory variable.

However, the above tests do not provide an indicator of the dynamic properties of the system and do not measure the relative strength of the Granger-causal chain or the degree of exogeneity among the variables beyond the sample period. Therefore, the final step of the empirical analysis investigates the properties of the system estimating the generalized impulse response function for the model. The purpose of the investigation is to find how each variable responds to shocks in the other variables of the system. The generalized impulse responses provide an estimate of the response of a variable in the case of innovation in another variable. Plotting the generalized impulse response functions is a practical way to explore the response of a variable to a shock immediately or with various lags. Unlike the orthogonalized impulse response functions obtained using the Cholesky factorization, the generalized impulse response functions are unique and invariant to the ordering of the variables in the VAR (Koop, Pesaran and Potter, 1996; Pesaran and Shin 1997).

The empirical analysis has been carried out using monthly data for the period 1989:1 to 2007:6 for Greece. The output variable is the industrial production (IP) (a measure of output), the real interest rates (RTB) are the 12-month rate minus the annual inflation rate as it is estimated employing the consumer price index, the oil price (OIL) is the consumer price index for fuels, the real stock (RSP) is the deflated stock index. All variables are not seasonally adjusted and except TB are expressed in logarithms (LIP, LRTB, LOIL, LRSP). All data are taken from the Bulletin of Conjectural Indicators of the Bank of Greece.

Empirical Results

The time series properties of all the variables used in the empirical analysis were evaluated employing standard unit-root tests. In particular, the augmented Dickey-Fuller (ADF), Phillips-Perron (PP), and the Kwiatkowski et al. (KPSS) test are used. All these tests suggested that industrial production, real interest rate, oil prices and real stock prices are non-stationary, while their first differences are stationary.[5]

Then, the existence of a long-run relationship among real interest rate, oil prices, real stock and industrial production is examined. To test for cointegration the Johansen maximum likelihood approach is used utilizing both the maximum eigenvalue and trace statistic. In estimating the long-run equation first a VAR system is constructed including four endogenous variables, industrial production, real interest rate, oil prices and real stock prices and seasonal dummies to account for seasonality. To determine the lag length of the VAR three versions of system are initially estimated: a twelve, a ten and an eight-lag version. Subsequently, an Akaike Information criterion (AIC), a Schwarz Bayesian Criterion (SBC) and a likelihood ratio test (Sims' test) are used to test that all three specifications are statistically equivalent. All tests reject the null hypothesis that all the versions of VAR are equivalent. In particular, the tests suggest that VAR=10 should be used in the estimation procedure of cointegration to avoid over-parameterization of the estimated models.

In addition, the estimated VAR system passes a series of diagnostic tests including serial correlation based on the inspection of the autocorrelation functions of the residuals as well as the reported Lagrange multiplier. In addition, it was tested whether the estimated regression equations were stable through-out the sample using the CUSUM and CUSUMSQ tests on structural stability of the estimated relations. The estimated statistics support the structural stability of the estimated regressions for both specifications.

Table 1. Johansen and Juselius Cointegration Test VAR=10, Industrial production, Real interest rate, oil prices, real stock prices 1989:1-2007:6

VAR=10 Variables: LIP, RTB, LOIL, LSR				
Maximum Eigenvalues				
Null	**Alternative**	**Eigenvalue**	**Critical Values**	
			95%	**90%**
r=0	r=1	29.94	27.07	32.24
r<=1	r=2	12.64	20.97	25.52
r<=2	r=3	7.59	14.07	18.63
r<=3	r=4	2.98	3.76	6.65
Trace Statistic				
Null	**Alternative**	**Trace**	**Critical Values**	
			95%	**90%**
r=0	r>=1	53.17	47.21	54.46
r<=1	r>=2	23.23	29.68	35.65
r<=2	r>=3	10.58	15.41	20.04
r<=3	r>=4	2.99	3.76	6.65

Note: r indicates the number of cointegrating relationships. Maximum eigenvalue and trace test statistics are compared with the critical values from Johansen and Juselius (1990).

5 The results on unit root tests are available from the author upon request.

The estimation procedure assumes unrestricted intercepts and no trends. Table 1 summarizes the results of cointegration analysis. The two test-statistics give similar results. Both tests provide evidence to reject the null of zero cointegrating vectors at 5% level. On the basis of the results the existence of one long-run relationship finds statistical support in Greece over the period under examination.

Having verified that the variables are cointegrated we investigate whether all the variables, industrial production, real interest rate, oil prices and real stock prices enter into the cointegrating vector in a statistically significant way. The results suggest that all variables enter in a statistically significant way into the cointegrating vector. The estimated cointegration relationship can be rewritten as where standard errors are reported in the parentheses:

$$LIP = -\,0.006RTB - 0.081LOIL + 0.083LRSP + 4.671$$
$$(0.003)\ (0.032)\ (0.011)$$

From the above estimated equation three main points can be concluded. First, a downward shock to real interest rate, due for example to changes in the monetary policy, leads to higher economic activity. Second, an upward shock in oil prices leads to lower industrial production. From a theoretical point of view various explanations have been offered to account for the inverse relationship between oil price changes and aggregate economic activity. Rising oil prices are an indication of an increased scarcity of labor. Since an important factor of production is scarce, output and labor productivity are reduced. In addition, as consumers expect a temporary rise in energy prices they may decide to save less causing a fall in real balances (Cologni and Manera, 2007). Finally, an upward movement of real stock prices is related to higher industrial production. The choice of the dependent variable is arbitrary since the estimated long-run equation based on the economic theory does not imply any causality among the variables. The causality channels are examined employing the VECM.

Having verified that the variables are cointegrated, vector error-correction models can be applied. The lagged residuals from the cointegrating regression with the appropriate number of lags are included in the Granger-causality test structure. The lag length structure depends on the restricted error-correction models. The restricted error-correction models pass a series of diagnostic tests including serial correlation based on the inspection of the autocorrelation functions of the residuals as well as the reported Lagrange multiplier.

The error-correction term measures the proportion by which the long-term imbalance in the dependent variable is corrected in each short-run period. The size and the statistical significance of the error-correction term measures the extent to which each dependent variable has the tendency to return to its long-run equilibrium.

Estimates of the parameters show that the error-correction term measuring the long-run disequilibrium is significant for all the equations at different levels of significance. This implies that all the variables have a tendency to restore equilibrium and take the brunt of any shock to the system. Specifically, the t-tests for the error-correction terms indicate that industrial production, real interest rate, oil prices and real stock prices are not weakly exogenous variable.

Table 2. Summary of Tests for Weak Exogeneity and Granger Causality of Variables Based on Vector Error-Correction Models

Equations	Test of Restrictions				
	Short-run Dynamics Non-Causality				Weak Exogeneity, (t-statistic for ECT coefficient)
	ΔLIP	ΔRTB	ΔLOIL	ΔLRSP	
ΔLIP		4.16	5.06	10.10	-1.74*
ΔRTB	17.68**		7.28	13.77	-3.06***
ΔLOIL	11.59	8.32		6.21	-2.18**
ΔLRSP	11.22	17.97**	9.27		1.71*

Note: The lagged ECT is derived by normalizing the cointegrating vector on industrial production. The statistics reported are distributed as chi-square distribution with degrees of freedom the number of restrictions. In the short-run dynamics asterisks indicate rejection of the H_0 that there is short-run non-causal relationship between the two variables. Asterisks indicate rejection of the null hypothesis that the estimated coefficient is equal to zero ***, **, * indicate significance at the 1, 5 and 10 percent levels.

In the short-run dynamics (Granger-causality in the strict sense), the Wald-tests indicate industrial production is exogenous hence is not affect by any variable in the short-run. Contrary stock prices are associated with the real interest rates.

Having verified the long-run relationship, the exogeneity and endogeneity, as well as the direction of causality among the four variables, the generalized variance decomposition analysis and the estimation of the generalized impulse response functions are used to investigate the dynamic properties of the system.

The generalized variance decomposition results for the four variables provide an indication of the relative strength of the Granger-causal relationship as well as a quantitative measure of the dynamic interactions between the variables. The results are presented in Table 3. The reported numbers indicate the percentage of the forecast error in each variable that can be attributed to innovations in other variables at five different time horizons: one year ahead (the short-run period), two or four years ahead (the medium-run period), eight or twelve years ahead (the long-run period). The results of generalized variance decomposition analysis and generalized impulse response function provide the same conclusions regardless of the order of decomposition since their estimation is independent on the ordering.

The analysis of generalized variance decomposition tends to suggest that each of the variables used in the empirical analysis can be explained by the disturbances in other variables after some periods. Specifically, for the industrial production variable own shocks account for 96.7% of the forecast variance for the six months. After a year (12 months, medium run), real interest rates, oil prices, and real stock prices account for 1.9%, 4.6%, and 10.1% of the industrial production forecast error variance, respectively. In the long-run the magnitudes of the explained variability of industrial production remain the same as in the medium-run period. For the interest rates in the first six months 92.9% of the variability in interest rates changes is explained by its own innovations, while after one year 76.1% of the variability is explained by its own innovations, 9.5% of the variability is explained by innovations in the industrial production, 5.4% in oil price changes and 9.0% in real stock prices. Finally, after two years, in the long-run period shocks to real interest rates, oil prices, industrial production and real stock prices account for approximately 65%, 6.2%, 13.2% and 15.4% of the variation in the 12-month real interest rate.

Table 3. Variance Decomposition Analysis of the Output Specification Generalized Variance Decomposition of Industrial Production

Period	Industrial production shock	Real interest rate shock	Oil prices shock	Real stock prices shock
1	100.0	0.0	0.0	0.0
6	96.7	1.6	0.7	0.9
12	83.4	1.9	4.6	10.1
18	81.7	1.6	4.8	11.9
24	79.3	1.4	5.1	14.2
Generalized Variance Decomposition of Real Interest Rate				
Period	**Industrial production shock**	**Real interest rate shock**	**Oil prices shock**	**Real stock prices shock**
1	0.0	100.0	0.0	0.0
6	2.4	92.9	0.2	4.6
12	9.5	76.1	5.4	9.0
18	11.5	69.8	6.0	12.7
24	13.2	65.2	6.2	15.4
Generalized Variance Decomposition of Oil Prices				
Period	**Industrial production shock**	**Real interest rate shock**	**Oil prices shock**	**Real stock prices shock**
1	0.1	8.2	91.7	0.0
6	5.0	6.1	88.5	0.3
12	16.6	6.1	74.9	2.4
18	20.7	7.3	67.4	4.5
24	23.1	7.9	62.8	6.2
Generalized Variance Decomposition of Real Stock Prices				
Period	**Industrial production shock**	**Real interest rate shock**	**Oil prices shock**	**Real stock prices shock**
1	0.1	0.5	1.1	98.4
6	5.7	2.3	8.0	84.0
12	9.6	2.4	6.3	81.8
18	14.5	2.5	5.4	77.6
24	18.0	2.7	4.9	74.4

Table 3 also presents the generalized variance decomposition for oil prices. In the first six months 88.5% of oil price changes variability comes from movements in itself. After a year more almost 75% of the variability is explained by its own innovations, while 16.6%, 6.1% and 2.4% are attributed to industrial production, real interest rate changes and real stock price changes. Finally, Table 3 presents the variance decomposition of real stock prices changes. In the first six months 84% of the real stock prices variability is attributed to shocks in itself, while 5.7% is due to changes in industrial production, 2.3% to real interest rate changes and 8.0% to oil price shocks. In the long-run period real stock returns variations are still mainly due to its own changes (74.4%), while 18% are attributed to industrial production changes, 2.7% to real interest rates and 4.9% to oil price changes.

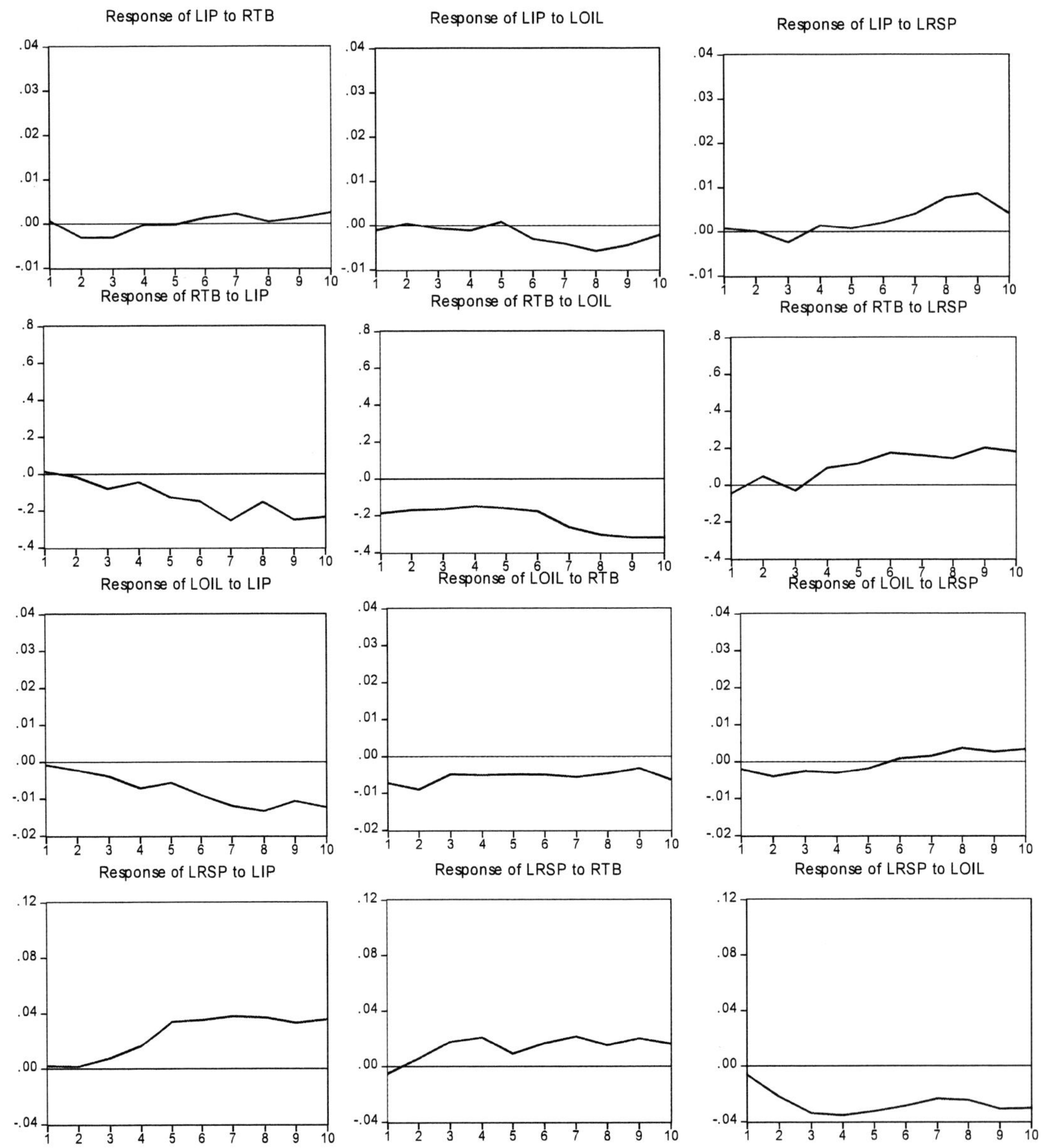

Figure 1. Response to Generalozed One S.D. Innovations.

An alternative way to obtain the information regarding the relationships among the four variables included in the variance decomposition analysis is through the generalized impulse response functions. Figure 1 presents the accumulated generalized impulse response functions for industrial production, real interest rates and real stock prices from one-standard deviation shock to industrial production, real interest rate, oil prices and real stock prices.[6]

Some comments for these graphs are in order. The graphs of Figure 1, first row, show the accumulated responses of industrial production, to a one-standard deviation shock on real interest rate, oil prices and real stock prices respectively. Real interest rate shocks (first row first graph) have a negative impact on industrial production as expected although this happens, as in Ferderer (1996) and Sardorsky (1999), only after an initial small positive impact. An oil price shock (first row, second graph) has an immediate negative effect on industrial production, as predicted by theory. This shock becomes higher in magnitude after the first six months. The response of industrial production to shocks in real stock prices is small in magnitude and becomes positive after the first four periods (first row, third graph). So as Lee (1992) we can conclude that the stock market rationally signal (or lead) changes in real activity but contrary to previous empirical findings for other industrial countries this signal is weak in the initial periods.

The graphs of Figure 1, second row, show the accumulated responses of real interest rates, to a one-standard deviation shock on industrial production, oil prices and real stock prices respectively. The graphs, of Figure 1, third row, show the accumulated responses of oil prices, to a one-standard deviation shock on industrial production, interest rates and real stock prices respectively.

Finally, the graphs of Figure 1, fourth row, show the accumulated responses of real stock prices to a one-standard deviation shock on industrial production, interest rates and oil prices respectively. An oil price shock has a negative impact on real stock prices (last row, fourth graph). As oil price changes affect economic activity in a negative way, and this affects the earnings of the companies. Higher production costs due to higher oil prices will cause earnings to decline. An efficient stock market will react with an immediate decline in stock prices (Sadorsky, 1999). Thus, oil price shocks depress real stock returns and increases in oil prices cause earnings to decline. The graph (last row, first graph) shows that the response of stock prices to shocks in industrial production is positive but accumulatively small in magnitude. So we can conclude that partially the performance of economic activity is transmitted in the stock market's performance.

Conclusion

In this paper the dynamic interactions among industrial production, real interest rates, oil prices and real stock prices are analyzed for Greece over the period 1989:1 to 2007:6. The VAR analysis employed, based on innovation accounting, allows the study of the dynamic interrelations among the variables without imposing a priori restrictions and reveals the different channels of causality among the variables. .

The evidence of cointegration implies that Granger-causality must exist among these variables either unidirectional or bi-directional. The paper through the use of vector error-

[6] The figure for each variable of its own innovation is omitted.

correction model detects the direction of Granger-causality, within sample Granger-exogeneity or endogeneity of each variable and distinguishes between short-term and long-term Granger causality. The results of the empirical model using the error-correction model estimation suggest that all the variables are not weakly exogenous. In addition, our findings indicate that in the short-run industrial production is exogenous while real stock prices are associated with the real interest rates.

From the empirical analysis of generalized impulse response functions significant conclusions can be drawn about the way in which oil price movements affect economic activity. Oil prices play important role in affecting economic activity. Oil price shocks explain a significant proportion of the fluctuations in output growth. Oil price shocks have an immediate negative effect on industrial production. In addition, impulse response functions show that oil prices are also important in explaining real stock price movements suggesting that a positive oil price shock depresses real stock returns.

Stock returns do rationally signal changes in real activity but this signal is very weak. Growth in industrial production responds positively to a real stock return shock implying that an increase in real stock returns leads to higher level of industrial production. However, the magnitude of this effect is small in size indicating that the signal from the stock market is partially transmitted to the industrial activity. Finally, real stock returns respond negatively to interest rate shocks. Interest rates and growth in industrial production are negatively associated, which suggests that a rise in interest rates rise is likely be associated with a lower growth in industrial production.

References

Burbridge, J. and A. Harrison. (1984). 'Testing for the Effects of Oil-Price Rises Using Vector Autoregressions', *International Economic Review*, Vol. 25, No.1, pp. 459-484.

Cologni, A. and Manera, M. (2007). 'Oil Prices, Inflation and Interest Ratesin a Structural Cointegrated VAR Model for the G-7 Countries', *Energy Economics*, forthcoming..

Chen, N.F. Roll, R. and Ross, S.A. (1986). 'Economic Forces and the Stock Market', *Journal of Business* ,Vol. 59, pp.383-403.

Choi, I. (1992). 'Effects of Data Aggregation on the Power of Tests for a Unit Root', *Economics Letters*, Vol. 40, pp. 397-401.

Dickey, D.A. and Fuller, W.A (1979). 'Distributions of the Estimators for Autoregressive Time Series with a Unit Root', *Journal of the American Statistical Association*, Vol. 74, pp. 427-31.

Dickey, D.A. and Fuller, W.A. (1981). 'The Likelihood Ratio Statistics for Autoregressive Time Series with a Unit Root', *Econometrica*, Vol. 49, pp. 1057-72.

Engle, R.F. and Granger, C.W.J. (1987). 'Cointegration and Error-Correction: Representation, Estimation and Testing', *Econometrica*, Vol. 55, pp. 251-76.

Ferderer, P.J. (1996). 'Oil Price Volatility and the Macroeconomy', *Journal of Macroeconomics*, Vol. 18, No. 1. pp. 1-26.

Gisser, M., and Goodwin, T.H. (1986). 'Crude Oil and the Macroeconomy: Tests of Some Popular Notions', *Journal of Money, Credit and Banking*, Vol. 18, No.1, pp. 95-103.

Guo, H. and. Kliesen, K (2005). 'Oil Price Volatility and U.S. Macroeconomic Activity', Federal Reserve Bank of St. Luis, *Review*, Vol. 87, No.6, pp. 669-683.

Hamilton, J.D. (1983). 'Oil and the Macroeconomy since World War II', *Journal of Political Economy*, Vol. 92, No. 2, pp. 228-248.

Hamilton J.D. (1996), 'This is What Happened to the Oil Price-Macroeconomy Relationship', *Journal of Monetary Economics*, Vol. 38, No.2, pp. 215-220.

Haug, R.D., R.W. Masulis, and Stoll, H.R. (1996). 'Energy Shocks and Financial Markets', *Journal of Futures Markets,* Vol. 16, No. 1, pp. 1- 27.

Johansen, S. (1988). 'Statistical and hypothesis Testing of Cointegration Vectors', *Journal of Economic Dynamics and Control*, Vol. 12, pp. 231-254.

Johansen, S. (1992). 'Cointegration in Partial Systems and the Efficiency of Single Equation Analysis', *Journal of Econometrics*, Vol. 52, pp. 389-402.

Johansen, S. and. Juselius, K. (1990). 'Maximum Likelihood Estimation and Inference on Cointegration - with Applications to the Demand for Money', *Oxford Bulletin of Economics*, Vol. 52, pp. 169-210.

Johansen, S. and Juselius, K. (1992). 'Testing Structural Hypotheses in a Multivariate Cointegration Analysis at the Purchasing Power Parity and the Uncovered Interest Parity for the UK', *Journal of Econometrics*, Vol. 53, pp. 211-244.

Jones, C.M. and Kaul, G. (1996). 'Oil and Stock Markets', *Journal of Finance*, Vol. 51, No. 2, pp. 463-491.

Jones, D.W., Paul P.N. and .Paik, P.I (2004). 'Oil Price Shocks and the Macroeoconomy: What Has Been Learned Since 1996?', *The Energy Journal*, Vol. 25, No. 2, pp. 1-32.

Kilian, L. and Park, C. (2007). 'The Impact of Oil Price Shocks on the U.S. Stock Market', *CEPR*, Discussion Paper, No. 6166.

Koop, G., M.H. Pesaran, and Potter, S.M. (1996). 'Impulse Response Analysis in Nonlinear Multivariate Models', *Journal of Econometrics*, Vol. 74, pp. 119-147.

Kwiatkowski, D., P.C.B. Phillips, Schmidt, P.and Shin, Y. (1992). 'Testing the Null Hypothesis of Stationarity Against the Alternative of a Unit Root', *Journal of Econometrics*, Vol. 54, pp. 159-78.

Lee, B. (1992). 'Causal Relationships Among Stock Returns, Interest Rates, Real Activity, and Inflation', *Journal of Finance*, Vol. 38, No.4, pp. 1591-1603.

Miguel, C., de, B. Manzano and Martin-Moreno, J.M. (2003). 'Oil Price Shocks and Aggregate Fluctuations', *Energy Journal* ,Vol.24, No.2, pp.47-61.

Mork, K.A (1989). 'Oil and the Macroeconomy When Prices Go Up and Down: An Extension of Hamilton's Results', *Journal of Political Economy*, Vol. 97, No. 3, pp. 740-744.

Mork, K.A., Olsen, O. and Mysen, H.T. (1994). 'Macroeconomic Responses to Oil Price Increases and Decreases in Seven OECD Countries', *Energy Journal,* Vol.15, No.4, pp.19-35.

Olsen, O. (1998). 'The Productivity Slowdown, the Oil Shocks, and the Real Cycle', *The Journal of Economic Perspectives*, Vol. 2, pp. 43-70.

Papapetrou, E. (2001). 'Oil Price Shocks, Stock Market, Economic Activity and Employment in Greece', *Energy Economics*, Vol. 23, pp. 511-532.

Perron, P. (1988). 'Trends and Random Walks in Macroeconomic Time: Series Further Evidence from a New Approach', *Journal of Dynamic and Control*, Vol. 12, pp. 297-332.

Perron, P. (1990). 'Tests of Joint Hypotheses in Time Series Regression with a Unit Root', in G.F. Rhodes, T.B. Fomby (Eds.), *Advances in Econometrics: Co-Integration, Spurious Regression and Unit Roots*, Vol. 8, New York, pp. 10-20.

Pesaran, M.H. and Shin, Y. (1996). Cointegration and Speed of Convergence to Equilibrium. Journal of Econometrics, Vol .71, pp. 117-143.

Phillips, P.C.B. (1987). 'Time Series Regression with a Unit Root', *Econometrica*, Vol. 55, pp. 277-346.

Phillips, P.C.B. and Perron, P. (1988). 'Testing for a Unit Root in Time Series Regression', *Biometrica*, Vol. 75, pp. 335-346.

Sadorsky, P. (1999). 'Oil Price Shocks and Stock Market Activity', *Energy Economics*, Vol. 21, pp. 449-469.

In: New Research on Energy Economics
Editors: P.G. Caldwell and E.V. Taylor, pp. 183-196
ISBN: 978-1-60456-354-2

Chapter 7

MACROECONOMIC EFFECTS OF OIL PRICE SHOCK IN THE G7 ECONOMIES

Juncal Cuñado and Fernando Perez de Gracia
Universidad de Navarra, Spain

Abstract

This chapter analyzes the economic impact of oil prices in the G7 economies over the period 1957Q1 to 2007Q1. Using a variety of definitions for oil price movements, the results suggest the following. First, oil prices have a greater impact on inflation and economic growth rates when they are measured in local currencies and in real terms. Second we find evidence of an asymmetric effect of oil prices on both inflation and economic growth rates. Furthermore, the economic impact of oil prices is significantly lower for the period after the mid-1980s than for the first subsample. Finally, the economic impact of this variable differs for the countries analyzed in this chapter.

Introduction

Historical evidence indicates that oil price movements have significant effects on macroeconomic variables. Many empirical studies found evidence that oil price shocks explain real GDP losses (see, for example, the seminals papers by Rasche and Tatom, 1977; Darby, 1982; Hamilton, 1983; Burbidge and Harrison, 1984 and Mork, 1989 among many others), reduce potential output (see, for example, Barro, 1984 and Brown and Yücel, 1999), increase the inflation rate (see, Hooker, 2002 and Cunado and Perez de Gracia, 2003,2005) and increase the unemployment rate (see, for example, Carruth et al., 1998).[1] Other theoretical studies such as Rotemberg and Woodford (1996), Finn (2000) and, recently, Blanchard and Gali (2007) try to incorporate oil price in alternative models specification in order to explain the macroeconomics effects of oil price shocks.

[1] For example, Brown and Yücel (2002), Jones et al. (2004) and recently Hamilton (2005) contain excellent surveys of the literature on the macroeconomic effects of oil price movements.

Before mid-1980s, there is a significant empirical evidence that oil shocks depressed economic activity. However, after 1985 the relationship between oil price and economic activity breaks down. Many researchers considered that instead of an symmetric (linear) effect from oil to economic growth (i.e., an increase in price of oil reduces GDP growth (Hamilton, 1983)) oil shocks may exert an asymmetric (non-linear) effect: oil price increases reduce output growth but oil price decreases do not increase output growth (Hooker, 1996).

In this chapter we focus on the macroeconomic effects of oil shocks for G7 countries using quarterly data for the period 1957Q1 to 2007Q1 taking into account the potential break in mid-1980s. We use alternative transformations of oil price data: symmetric versus asymmetric, each of one suggesting a different channel through which oil prices may affect industrial production levels. Initially, we examine the relationship between oil price - inflation and oil price - economic activity for the period 1957 to 2007. Finally, the possible structural break in mid-1980s is also investigated, proposing two different subsamples for the empirical analysis.

This chapter is structured as follows. Section 2 presents alternative variables for oil price movements. Section 3 presents the empirical analysis in the G7 economies for the full sample. Section 4 provides empirical evidence taking into account the potential break in the relationship between macroeconomy and oil shocks. Section 5 concludes with a summary of the chapter.

Alternative Measures for Oil Price Shocks

In this section we present alternative measures for oil shocks already used in many studies. Before the definition of the oil price variables, we provide previously some information about the data set.

The quarterly data used in this study are obtained from International Financial Statistics (International Monetary Fund) and cover the period 1957Q1-2007Q1. The countries included in the study are the G7 economies: Canada, France, Germany, Italy, Japan, the UK and the US. The employed variables are:

- Real oil price: the producer price index for crude oil is deflated using the consumer price index of each of the countries. We use the exchange rate for each G7 economy (except the US) against the $US. Originally, oil prices are expressed in $US and they are converted to local currencies using the average market exchange rates for the quarter.
- Inflation rates: calculated from Consumer Price Index (CPI).
- Economic activity: proxied by Industrial Production Index seasonally adjusted.

The choice of oil price variables is quite relevant and, as we shall show later, important. Previous studies on the macroeconomic effects of oil price movements use different definitions for oil shocks: for example, $US nominal oil price (Hamilton, 1996), $US real oil price (Hooker, 1999), nominal oil price in local currency or real oil price in local currency (Mork et al., 1994). National oil prices have been influenced by price controls, varying taxes on petroleum products, exchange rate and national price index variations. All the differential characteristics which influence the effective oil price that each of the countries face raise great difficulties in measuring the appropriate oil price variable for each country.

In this chapter we will measure oil prices in different ways: in $US, in local currency and nominal terms and in local currency and real terms. We believe this analysis will help us to analyze different channels through which oil prices may affect economy (i.e., net oil price, inflation rates, exchange rates, etc.).

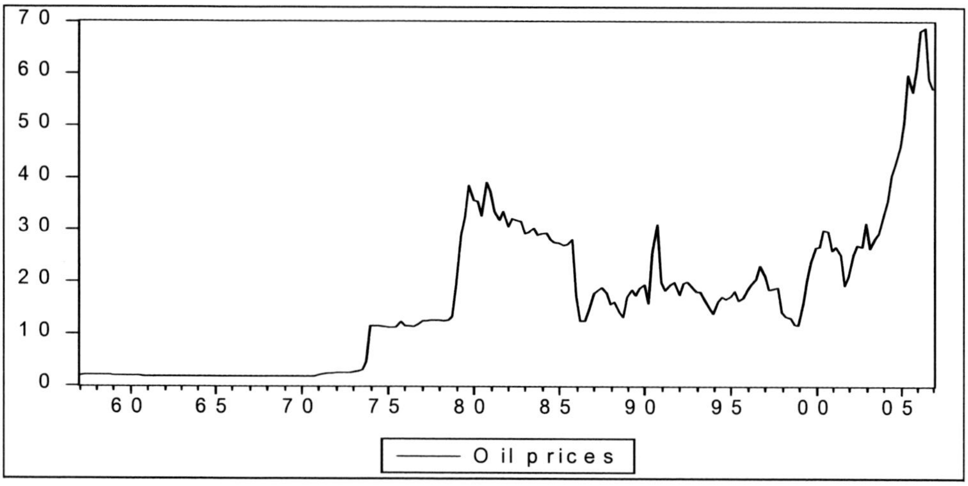

Figure 1. Nominal oil price in $US.

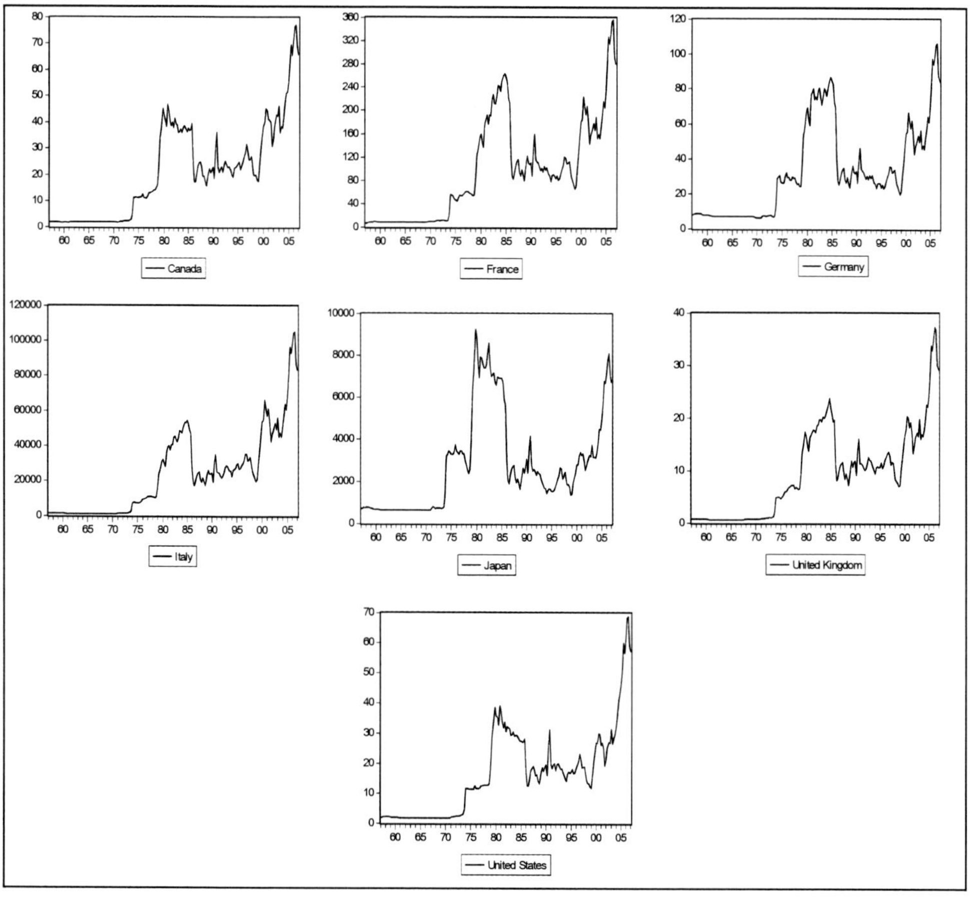

Figure 2. Nominal oil prices in local currencies.

We initially present the evolution of nominal oil price for the period 1957 to 2007. As we can see in Figure 1, the nominal oil price in $US was very stable from the late 1950s to the early 1970s. During the period 1957 to 1972, the oil price was directly influenced by the oil production quotas. From 1973 to mid-1980s, the price of oil rose dramatically due to the Yom Kippur War in 1973 and 1974 and the Iran-Iraq war in 1979 and 1980. Since mid-1980s to 2007, the variability in oil price has been much greater than in the past. A first look at Figure 1 suggests the existence of a potential break in the oil price movements around mid-1980s.

Figure 2 shows for each of the G7 countries, the evolution of nominal oil prices expressed in local currency of each economy. In all the series we observe the effects of the fourth main negative oil increases located in 1978-1979, 1990, 1999-2000 and 2005-2006. Figure 3 shows the real oil price for each of the G7 economies also expressed in local currency. Recall from Figure 3 that the higher price of oil changes were located over the period 1980-1985. Further, since mid-1980s it is worth mentioning the fall in oil prices due to the market collapsed in mid-1980s and the generalized nominal price decreases.

Table 1 presents the correlation coefficients among nominal oil prices expressed in local currency. The Table shows a high degree of correlation both among the nominal price of oil for all G7 countries - in no case, the coefficient is lower than 0.73-. Table 2 also presents the correlation coefficients among real price of oil expressed in local currency. The correlation coefficients are in all cases greater than 0.92 except in the case of France versus Japan and Japan versus the US. The initial evidence suggest that nominal and real oil prices tend to move synchronized over the sample period for most of the G7 economies.

Table 1. Correlation coefficients among local nominal oil prices

	France	Germany	Italy	Japan	UK	US
Canada	0.97	0.94	0.97	0.82	0.99	0.99
France		0.98	0.97	0.85	0.99	0.95
Germany			0.91	0.94	0.96	0.94
Italy				0.73	0.83	0.95
Japan					0.83	0.85
UK						0.98

Note: nominal oil prices are expressed in local currency using exchange rates againts $US.

Table 2. Correlation coefficients among local real oil prices

	France	Germany	Italy	Japan	UK	US
Canada	0.95	0.94	0.96	0.87	0.94	0.99
France		0.99	0.99	0.90	0.97	0.94
Germany			0.99	0.92	0.97	0.94
Italy				0.93	0.98	0.96
Japan					0.92	0.90
UK						0.95

Note: real local oil prices are defined as oil prices expressed in local currency and in real terms.

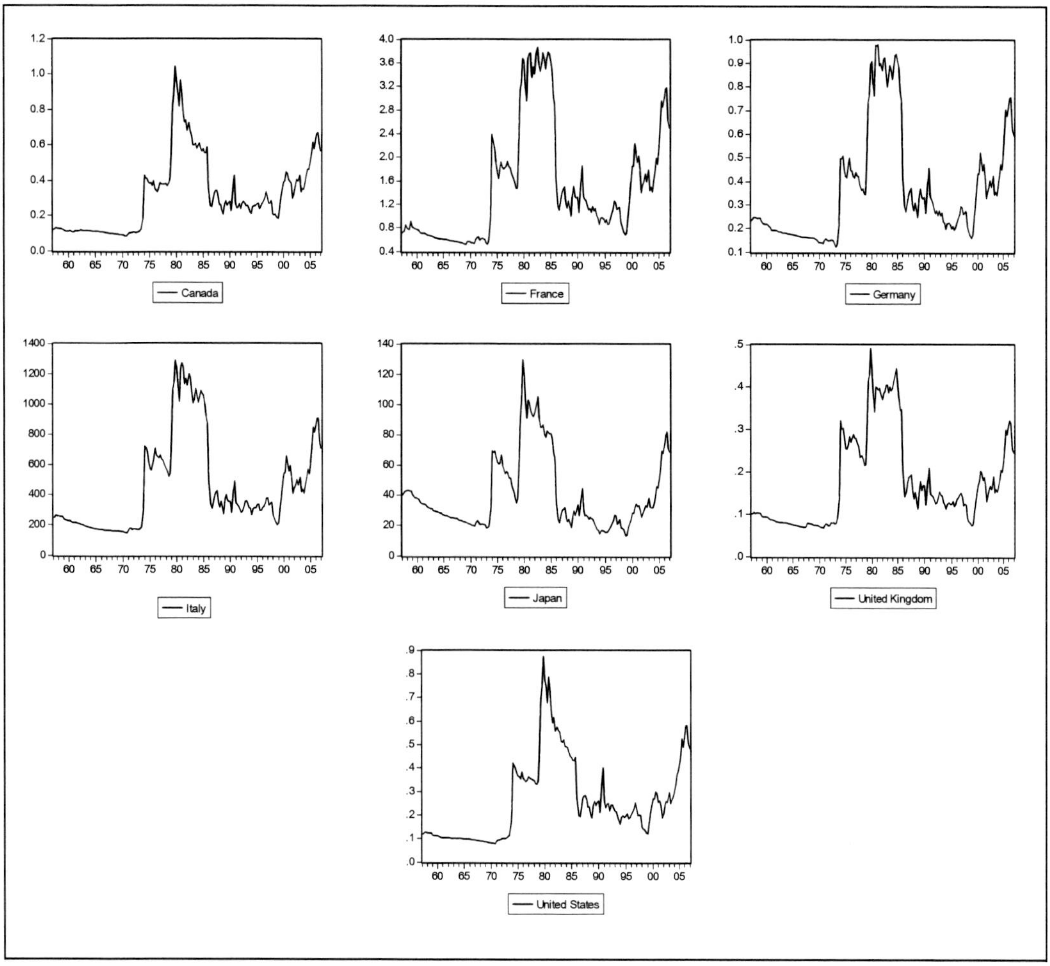

Figure 3. Real oil prices in local currencies.

As we mentioned previously in the Introduction section, the oil price tend to present a differential behavior before and after mid-1980s. This potential break have led researchers to explore different oil-output specifications in order to reestablish the relationship between these variables (see, for example, Mork, 1989; Lee et al., 1995; Hamilton, 1996 and Cunado and Perez de Gracia, 2003, 2005 among others).

Following previous studies, we propose alternative oil shocks that allow for a symmetric and an asymmetric relationship between oil and macroeconomic variables. These oil price proxies are:

$\Delta noil_t$: quarterly changes of nominal oil prices, that is, the conventional first difference transformation of oil price variables (in logs):

$$\Delta \text{noil}_t = \ln(\text{noil}_t) - \ln(\text{noil}_{t-1}) \quad (1)$$

where $noil_t$ is the nominal oil price in period t in \$US or in local currency, as defined above. A significant relationship between this variable and economic activity would lead to a linear (symmetric) oil-output relationship.

$\Delta roil_t$: quarterly changes of real oil prices:

$$\Delta \text{roil}_t = \ln(\text{roil}_t) - \ln(\text{roil}_{t-1}), \tag{2}$$

where $roil_t$ is the real oil price in period t in \$US or in local currency, as defined above.

We also used the following three asymmetric proxies:

- $\Delta noil_t^+$: nominal oil price increases, defined as

$$\Delta \text{noil}_t^+ = \max(0, \Delta \text{noil}_t). \tag{3}$$

In this case, we treat in a different way oil price increases and decreases, that is, we separate oil price changes into negative and positive changes in a believe that oil price increases may have a significant effect on macroeconomic variables even though this might not occur for oil price decreases.

$\Delta roil_t^+$: real oil price increases,

$$\Delta \text{roil}_t^+ = \max(0, \Delta \text{roil}_t). \tag{4}$$

$NOPI_t$: net oil price increases (expressed in real terms) defined as the quarterly percentage change in real oil price levels from the past 12 quarters' high if that is positive and zero otherwise. This variable is proposed by Hamilton (1996), who argues that if one wants a measure of how unsettling an increase in the price of oil is likely to be for the spending decisions of consumers and firms, it seems more appropriate to compare the current price of oil with where it has been over the previous years rather than during the previous quarter alone. Hamilton thus proposes to use the amount by which the log oil price in quarter t exceeds its maximum value over the previous periods; if oil prices are lower than they have been at some point during the most recent years, no oil shock is said to have occurred. That is,

$$\text{NOPI}_t = \max\left[0, (\ln(\text{oil}_t) - \ln(\max(\text{oil}_{t-1}, \ldots, \text{oil}_{t-12})))\right]. \tag{5}$$

With this variable, we aim to check for a causal relationship between "important" oil price increases and macroeconomic variables.

The alternative proxies for oil prices (i.e., nominal and real oil price increases, positive nominal and real oil price increases and NOPI) are plotted in Figure 4 and the correlation coefficients among them are reported in Table 3. As we can see in the Figure, the different oil price proxies detect quite well the three main oil shocks in the period 1957Q1-2007Q1. The first one takes place in 1978-1979 when the Iranian revolution disrupted oil supplies and the price rose from \$20 to \$30. A second one followed Iraq´s invasion of Kuwait in 1990, when prices went from \$16 to \$26. Finally, prices grew up from \$12 to \$24 in 1999. However, we can also detect some differences among each of the variables. For example, we can observe that the variables $\Delta noil^+$ and $\Delta roil^+$ takes a much higher value after the increase in oil prices after the invasion of Kuwait (see 1990Q3) than the NOPI variable, a difference which is due to the decrease in oil prices occurred in 1990Q2.

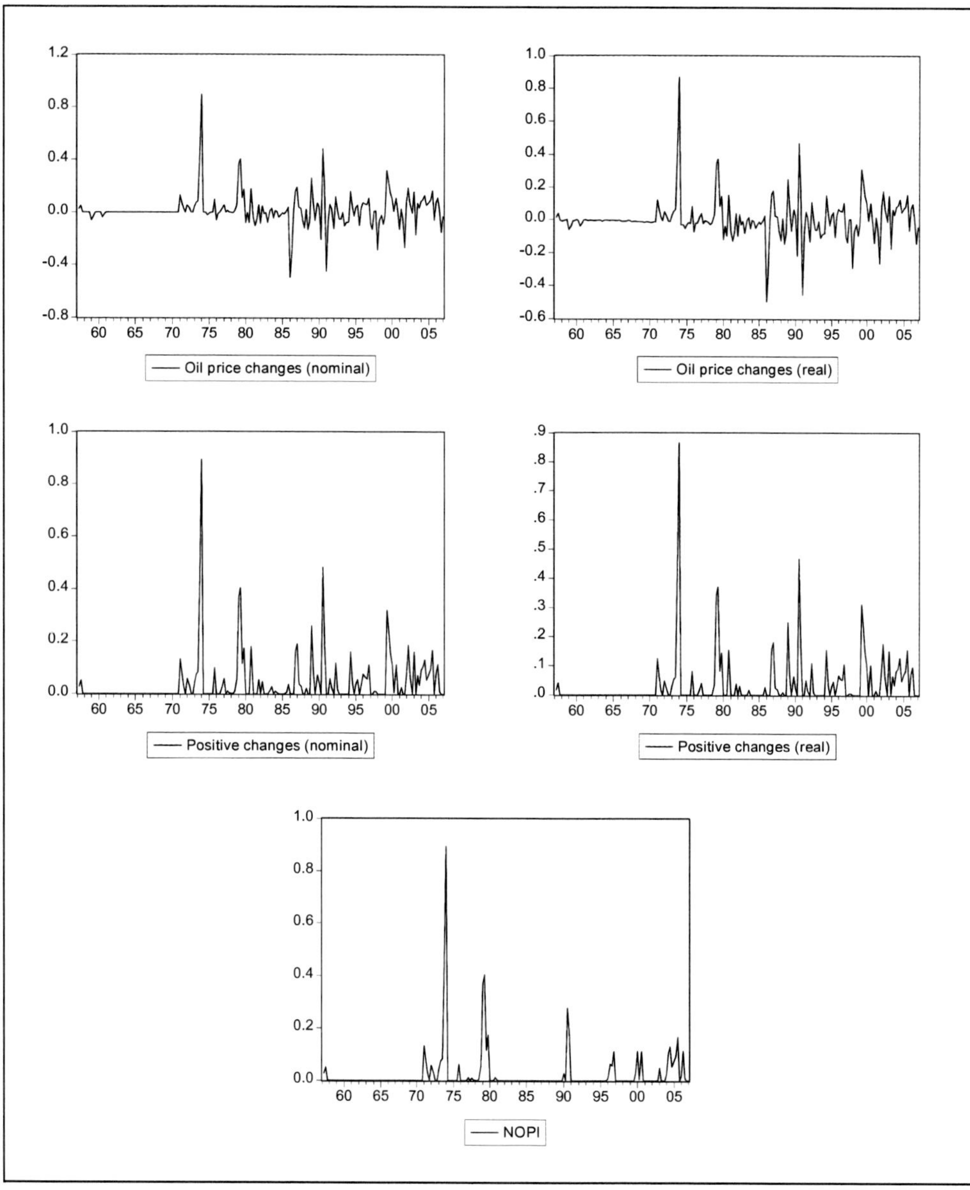

Figure 4. Alternative ways of measuring oil price shocks.

Table 3. Correlation coefficients among oil price proxies

	Δroil	**Δnoil^{+}**	**Δroil^{+}**	**NOPI**
Δnoil	0.99	0.86	0.85	0.73
Δroil		0.86	0.85	0.72
Δnoil^{+}			0.99	0.87
Δroil^{+}				0.87

Notes: Δnoil: nominal oil price changes; Δnoil^{+}: positive nominal oil price changes; Δroil: real oil price changes; Δroil^{+}: positive real oil price changes and NOPI: net real oil price increase. All these proxies have been constructed using the oil price variable defined in $US.

Economic Impact of Oil Price Shocks

In this section we examine the macroeconomic effects of oil shocks, by means of estimating the impact of oil prices on both economic activity and inflation rates for the G7 countries during the period 1957Q1-2007Q1.[2]

The estimation strategy is as follows. First, we check for stationarity in each variable. Second, we analyze whether a long-run relation exists between the series testing for bivariate cointegration between oil prices and both inflation and Industrial Production Index (IPI). Third, we study the short-run dynamic behavior between oil prices and macroeconomic variables checking for Granger causality. We also take into account for asymetries in the relationship between oil price shocks and economic growth and inflation rates.

As a first step of the empirical analysis, unit-root tests have been carried out for all of the variables: oil prices (both in nominal and real terms), inflation and IPI. The results (avalaible upon request) show that inflation, industrial production and oil prices may be considered integrated of order one (I(1)) variables in all countries. As all the variables exhibit a unit-root, we also tested for bivariate cointegration. The general result (also avalaible upon request) of this analysis is that there is not a cointegrating long-run relationship between oil prices and economic activity, which suggests that the impact of oil shocks on these variables is limited to the short-run.

In order to test for Granger causality from oil prices to economic growth rates and inflation rates, we specify a short-run relationship between the variables estimating the following model in which all the variables are stationary:

$$\Delta\pi_t = \alpha_0 + \sum_{i=1}^{k} \alpha_{1i}\Delta\pi_{t-i} + \sum_{i=1}^{k} \alpha_{2i} \left\{ \begin{matrix} \Delta \mathrm{noil}_{t-i} \\ \Delta \mathrm{noil}^{+}_{t-i} \\ \Delta \mathrm{roil}_{t-i} \\ \Delta \mathrm{roil}^{+}_{t-i} \\ \mathrm{NOPI}_{t-i} \end{matrix} \right\} + \varepsilon_t, \tag{6}$$

$$\Delta y_t = \alpha_0 + \sum_{i=1}^{k} \alpha_{1i}\Delta y_{t-i} + \sum_{i=1}^{k} \alpha_{2i} \left\{ \begin{matrix} \Delta \mathrm{noil}_{t-i} \\ \Delta \mathrm{noil}^{+}_{t-i} \\ \Delta \mathrm{roil}_{t-i} \\ \Delta \mathrm{roil}^{+}_{t-i} \\ \mathrm{NOPI}_{t-i} \end{matrix} \right\} + \varepsilon_t, \tag{7}$$

where Δ is the difference operator, k is the number of lags (chosen according to the AIC criteria), the dependent variable are both the first difference of inflation rates and the growth rate of production, and as explanatory variables we include lagged values of the dependent

2 Inflation rate (π) is calculated from Consumer Price Index (CPI) and economic activity (y) is proxied by the growth rate of Industrial Production Index.

variable and of each of the constructed proxies of oil price shocks (Δnoil, $\Delta noil^+$, Δroil, $\Delta roil^+$ and NOPI).[3]

Failing to reject the null hypothesis $\alpha_{21}=\alpha_{22}=...=\alpha_{2k}=0$ implies that oil price changes do not Granger cause inflation rates or economic growth. As far as the oil prices - inflation rates relationship is concerned, Table 4 presents the Granger causality tests from oil price shocks to inflation rates. The main results may be summarized as follows. First, when oil prices are defined in $US, we only find evidence of causality (at 5% level) for Italy and the US. However, the evidence of causality is higher when oil prices are measured in domestic currencies (in this case, the null hypothesis is rejected for Canada, Italy, Japan and the US at the 5% level) and even greater when they are measured in domestic currencies and real terms (we reject the null hypothesis for all countries except for Germany). Finally, the evidence of causality increases when we allow for asymmetric effects of oil shocks on inflation rates (except, again, for the case of Germany).

Table 4. Granger causality tests from oil prices to inflation rates (full sample, 1957Q1-2007Q1)

	Δnoil($)	Δnoil	Δroil	$\Delta noil^+$($)	$\Delta noil^+$	$\Delta roil^+$	NOPI
Canada	2.15*	2.69**	2.95**	2.41*	2.91**	2.85**	1.75
France	1.89	2.36*	2.52**	3.70***	4.00***	4.25***	4.18***
Germany	1.10	1.78	1.84	0.81	1.12	1.14	0.40
Italy	3.94***	6.02***	6.78***	5.00***	8.10***	7.53***	7.53***
Japan	2.34*	3.37**	3.47***	4.31***	4.74***	4.63***	4.83***
UK	2.42*	2.18*	2.73**	3.65***	3.50***	4.38***	5.40***
US	2.76**		2.79**	2.94**		2.82**	2.00**

Notes: Δnoil($): nominal oil price changes in $US; Δnoil: nominal oil price changes expressed in domestic currency; Δroil:real oil price changes expressed in domestic currency; Δnoil+($): positive nominal oil price changes in $US; Δnoil+: positive nominal oil price changes expressed in local currency; Δroil+: positive real oil price changes expressed in local currency; NOPI: net oil price increase with 12 quarters; inflation (π) is defined as the growth rate of CPI. The number of lags have been chosen according to the AIC criteria. The null hypothesis that lag values of oil price shocks are not significant in explaining changes in inflation rates is tested (see equation (6)). *, ** and *** indicate significant at 10, 5 and 1% respectively.

Table 5 shows the results of these Granger causality tests from oil price shocks to economic growth rates. In this case, the short-run relationship between these variables is weaker, since we only find evidence of causality for Canada, Japan, the UK and the US when oil prices are defined in $US. When oil prices are defined in domestic or local currencies, we can only reject the null hypothesis for Canada, the UK and the US. However, we find clear evidence of causality when we allow for an asymmetric relationship. In this case, we reject the null hypothesis for all the G7 economies.

[3] Recall that the defined proxy variables for oil price changes are stationary transformations of oil prices in $US and in each of the domestic currencies, which are I(1) variables.

Table 5. Granger causality tests from oil prices to economic growth rates (full sample, 1957Q1-2007Q1)

	Δnoil($)	**Δnoil**	**Δroil**	**Δnoil^{+}($)**	**Δnoil^{+}**	**Δroil^{+}**	**NOPI**
Canada	3.18**	2.81**	2.54**	5.65***	5.23***	5.16***	4.43***
France	1.57	1.55	1.52	3.55***	3.12**	2.93**	3.33**
Germany	1.70	1.65	1.58	2.69**	2.36*	2.26*	2.45**
Italy	1.47	1.73	1.59	3.34**	3.38**	3.04**	2.93**
Japan	2.84**	1.86	1.72	5.62***	4.97***	4.80***	4.68***
UK	3.96***	3.65***	3.61***	6.52***	5.68***	5.32***	7.71***
US	3.60**		3.25**	6.69***		6.60***	6.74***

Notes: Δnoil($): nominal oil price changes in $US; Δnoil: nominal oil price changes expressed in domestic currency; Δroil: real oil price changes expressed in domestic currency; Δnoil^{+}($): positive nominal oil price changes in $US; Δnoil^{+}: positive nominal oil price changes expressed in local currency; Δroil^{+}: positive real oil price changes expressed in local currency; NOPI: net oil price increase with 12 quarters; y is proxied by Industrial Production Index. The number of the lags included was determined according to the AIC criteria. The null hypothesis that lag values of oil price shocks are not significant in explaining economic growth rates is tested (see equation (7)). *, ** and *** indicates significant at 10, 5 and 1% respectively.

Potential Break in Mid-1980s

As we have shown previously, oil price dynamics presents a potential break in mid-1980s. Many empirical studies have documented that the impact of oil price shocks do not affect economic variables over time and tend to decay since mid-1980s (Hooker, 1996).

In this section we split the sample period into two subsamples: the first subsample runs from 1957Q1 to 1985Q4 and the second subsample runs from 1986Q1 to 2007Q1. In Tables 6 and 7 we display the results taken into account the potential break in 1985. As far as the inflation rates-oil price changes relationship is concerned, the main results suggest that there is clear evidence of a higher impact of different price of oil transformations in the first subsample than in the second subsample for France, Italy, the UK and the US (see Table 6).

Table 6. Granger causality tests from oil prices to inflation rates

	Δnoil($)	**Δnoil**	**Δroil**	**Δnoil^{+}($)**	**Δnoil^{+}**	**Δroil^{+}**	**NOPI**
First subsample (1957:Q1-1985:Q4)							
Canada	1.11	1.41	1.60	1.15	1.24	1.24	1.10
France	2.36*	2.35*	2.68**	2.54*	2.27*	2.37*	2.79**
Germany	0.59	0.95	0.96	0.70	0.72	0.68	0.64
Italy	3.53***	6.83***	7.82***	4.06***	6.87***	6.46***	5.15***
Japan	1.74	2.83**	2.76**	2.37*	2.79**	2.73**	2.80**
UK	4.96***	4.34***	5.48***	4.63***	4.23***	5.44***	4.89***
US	2.87**		2.92**	2.90**		2.66**	2.12*

Table 6. Continued

	Δnoil($)	Δnoil	Δroil	$\Delta noil^{+}$($)	$\Delta noil^{+}$	$\Delta roil^{+}$	NOPI
Second subsample (1986:Q1-2007:Q1)							
Canada	1.07	1.21	1.29	1.73	2.19*	2.25*	2.17*
France	1.47	1.82	1.83	1.69	2.72**	2.76**	1.48
Germany	1.71	2.00	2.05*	3.14**	3.23**	3.14**	2.22*
Italy	0.96	2.01	2.09*	1.02	1.63	1.67	0.30
Japan	3.54**	4.74***	4.83***	2.63**	3.63***	3.74***	0.74
UK	0.34	0.53	0.50	0.67	0.92	0.84	1.60
US	2.30*		2.33*	1.07		1.11	1.37

Notes: Δnoil($): nominal oil price changes in $US; Δnoil: nominal oil price changes expressed in domestic currency; Δroil: real oil price changes expressed in domestic currency; $\Delta noil^{+}$($): positive nominal oil price changes in $US; $\Delta noil^{+}$: positive nominal oil price changes expressed in local currency; $\Delta roil^{+}$: positive real oil price changes expressed in local currency; NOPI: net oil price increase with 12 quarters; inflation (π) is defined as the growth rate of CPI. The number of the lags included was determined according to the AIC criteria. The null hypothesis that lag values of oil price shocks are not significant in explaining economic growth rates is tested (see equation (6)). *, ** and *** indicates significant at 10, 5 and 1% respectively.

Table 7. Granger causality tests from oil prices to economic activity

	Δnoil($)	Δnoil	Δroil	$\Delta noil^{+}$($)	$\Delta noil^{+}$	$\Delta roil^{+}$	NOPI
First subsample (1957:Q1-1985:Q4)							
Canada	2.70**	2.60**	2.21*	3.68***	3.46**	3.49**	2.96**
France	3.37**	4.08***	3.67***	3.81***	4.07***	3.76***	3.73***
Germany	2.11*	2.56**	2.42*	2.58**	2.87**	2.77**	2.39*
Italy	2.83**	3.24**	2.75**	3.02**	3.11**	2.71**	2.67**
Japan	4.77***	3.76***	3.05**	5.54***	4.62***	4.18***	4.80***
UK	5.29***	4.60***	4.67***	5.33***	4.68***	4.49***	5.69***
US	5.19***		4.69***	6.05***		6.07***	5.21***
Second subsample (1986:Q1-2007:Q1)							
Canada	1.57	1.13	1.05	2.97**	2.17*	2.17*	3.19**
France	0.35	0.74	0.73	0.50	1.02	1.00	0.11
Germany	1.38	1.38	1.45	1.53	2.27*	2.24*	0.90
Italy	0.31	0.26	0.27	0.02	0.09	0.09	0.20
Japan	1.30	0.60	0.58	1.13	1.26	1.23	1.71
UK	1.51	1.53	1.39	1.30	1.03	0.89	2.20*
US	1.58		1.52	2.23*		2.22*	2.99**

Notes: Δnoil($): nominal oil price changes in $US; Δnoil: nominal oil price changes expressed in domestic currency; Δroil: real oil price changes expressed in domestic currency; $\Delta noil^{+}$($): positive nominal oil price changes in $US; $\Delta noil^{+}$: positive nominal oil price changes expressed in local currency; $\Delta roil^{+}$: positive real oil price changes expressed in local currency; NOPI: net oil price increase with 12 quarters; y is proxied by Industrial Production Index. The number of the lags included was determined according to the AIC criteria. The null hypothesis that lag values of oil price shocks are not significant in explaining economic growth rates is tested (see equation (7)). *, ** and *** indicates significant at 10, 5 and 1% respectively.

Table 7 presents the Granger causality tests from oil price variables to economic growth rates. In this case, we find that oil price shocks in the second subsample have a lower and non-significant effects on economic growth rates in all the G7 economies.

Conclusion

In this chapter we have documented the economic impact of oil price shocks on both inflation and economic growth rates in the G7 economies for the period 1957 to 2007. In order to detect the different channels through which oil prices may affect economy (i.e., net oil price, inflation rates, exchange rates, etc.), we use alternative proxies for oil price movements: nominal and real domestic oil price changes, nominal and real positive oil price changes and net oil price increases.

The main results are the following. First, when we analyze the whole period and measured oil prices in $US, we only find evidence of Granger causality (at 5% level) from oil prices to inflation rates for Italy and the US, and from oil prices to economic activity for Canada, Japan, the UK and the US. Second, when oil prices are measured in national currencies and real terms, we find causality from oil prices to inflation rates for all countries except for Germany. However, for the oil prices - economic activity relationship, we only find evidence of linear (symmetric) causality for Canada, the UK and the US. However, when we allow for a nonlinear (asymmetric) relationship between oil prices and macroeconomic variables, we reject the null hypothesis of non-causality for all countries (except for inflation rate in the case of Germany). Finally, when we split the sample period in mid-1980s, we conclude that the macroeconomic impact of oil prices are higher before 1985 than after that period.

As we have shown, since mid-1980s recent oil price shocks have lower impact on economic growth rates and inflation rates. This result obtained is also supported by other studies such as Roubini and Setser (2004), Rogoff (2006) and Blanchard and Gali (2007) among others. Roubini and Setser (2004) in a descriptive analysis conclude that the economic impact of oil price shocks were higher in the 1970s for many reasons: first, real oil price rose to a higher level in the 1970s; second, the speed of the change in price were also larger in the 1970s; third, oil price shocks in the 1970s were more persistent and a higher economic dependence on oil in the 1970s among many others. Similarly, Rogoff (2006) arguments different factors such as a greater energy efficiency, a greater concentration of oil consumption in final demand, better anchored monetary policy, deeper financial markets and more flexible labor markets. Finally, Blanchard and Gali (2007) also find a weaker response of inflation, GDP and employment to oil price shocks in recent years explained more flexible labor markets, improvements in monetary policy and a smaller share of oil in production.

Acknowledgement

Juncal Cunado gratefully acknowledges financial support from the Spanish Ministry of Science and Technology (SEJ2005-07657/ECON) and Fernando Perez de Gracia acknowledges research support from the Spanish Ministry of Science and Technology and

FEDER through grant SEJ2005-06302/ECON and from the Plan Especial de Investigacion de la Universidad de Navarra.

References

Barro, R. (1984). *Macroeconomics*. New York: John Wiley and Sons.

Blanchard, O., and Gali, J. (2007). The macroeconomic effects of oil price shocks: Why are the 2000s so different from the 1970s?, *Mimeo*.

Brown, S.P., and Yücel, M.K. (1999). Oil prices and US aggregate economic activity. A question of neutrality. *Economic and Financial Review. Federal Reserve Bank of Dallas, Second Quarter*, 16-23.

Brown, S.P., and Yücel, M.K. (2002). Energy prices and agregate economic activity: An interpretative survey. *The Quaterly Review of Economic and Finance*, 42, 193-208.

Burbidge, J., and Harrison, A., (1984). Testing for the effects of oil-price rises using vector autoregression. *International Economic Review*, 25, 459-484.

Carruth, A.A., Hooker, M.A., and Oswald, A.J. (1998). Unemployment equilibria and input prices: Theory and evidence from the United States. *Review of Economics and Statistic*, 80, 621-628.

Cunado, J., and Perez de Gracia, F. (2003). Does oil price shocks matter? Evidence for some European countries. *Energy Economics*, 25, 137-154

Cunado, J., and Perez de Gracia, F. (2005). Oil Prices, economic activity and inflation: Evidence for some Asian countries. *The Quarterly Review of Economics and Finance*, 45, 65-83.

Darby, M. (1982). The price of oil and world inflation and recesion. *American Economic Review*, 72, 738 – 751.

Finn, M. (2000). Perfect competition and the effects of energy price increases on economic activity. *Journal of Money Credit and Banking*, 32, 400-416.

Hamilton, J. (1983). Oil and the macroeconomy since World War II. *Journal of Political Economy*, 91, 593-617.

Hamilton, J. (1996). This is what happened to the oil price-macroeconomy relationship. *Journal of Monetary Economy*, 38, 215-220.

Hamilton, J. (2005). *Oil and the macroeconomy*. Prepared for Palgrave Dictionary of Economics. Forthcoming.

Hooker, M. (1996). What happened to the oil price-macroeconomy relationship?. *Journal of Monetary Economics*, 38, 195-213.

Hooker, M. (1999). Oil and the macroeconomy revisited. *Federal Reserve Board Working Paper* No. 99-43.

Hooker, M. (2002). Are oil shocks inflationary? Asymetric and nonlinear specifications versus change in regime. *Journal of Money, Credit and Banking*, 34, 540-561.

Jones, D., Leiby, P., and Paik, I. (2004). Oil price shocks and the macroeconomy: What has been learned since 1996. *The Energy Journal*, 25, 1–32.

Lee, K., Ni, S., and Ratti, R.A. (1995). Oil shocks and the macroeconomy: The role of price variability. *The Energy Journal*, 16, 39-56.

Mork, K. (1989). Oil and the macroeconomy when prices go up and down: An extension of Hamilton´s resuls. *Journal of Political Economy*, 97, 740-744.

Mork, K., Olsen, O., and Mysen, H.T. (1994). Macroeconomic responses to oil price increases and decreases in seven OECD countries. *Energy Journal,* 15, 15-38

Rasche, R. H., and Tatom, J.A. (1977). Energy resources and potential GNP. *Federal Reserve Bank of St. Louis Review,* 59, 10–24.

Rogoff, K. (2006). Oil and the global economy. *Presented at the International Energy Forum*, Riyath, November 19.

Rotemberg, J., and Woodford, M. (1996). Imperfect competition and the effects of energy price increases on economic activity. *Journal of Money, Credit and Banking,* 28, 549-577.

Roubini, N., and Setser, B. (2004). The effects of recent oil price shock on the US and global economy. *Mimeo.*

In: New Research on Energy Economics
Editors: P.G. Caldwell and E.V. Taylor, pp. 197-209
ISBN: 978-1-60456-354-2

Chapter 8

GASOLINE PRICES: NEW LEGISLATION AND PROPOSALS*

Carl E. Behrens and Carol Glover

Abstract

The high price of gasoline was an important consideration during the debate on major energy legislation, which ended August 8 as the President signed the Energy Policy Act of 2005, H.R. 6 (P.L. 109-58). However, prices continued to surge, spiking at the end of August when Hurricane Katrina shut down refining operations in the Gulf of Mexico. The continuing crisis renewed attention to some issues that were dropped or compromised in the debate over P.L. 109-58, as well as a number of initiatives to reduce the impact of high prices on consumers.

A large number of factors combined to put pressure on gasoline prices, including increased world demand for crude oil and U.S. refinery capacity inadequate to supply gasoline to a recovering national economy. The war and continued violence in Iraq added uncertainty and a threat of supply disruption that added pressure particularly to the commodity futures markets.

Among the issues that received new attention were vehicle fuel economy standards, leasing on the Outer Continental Shelf, and refinery "revitalization" provisions. The Gasoline for America's Security Act of 2005, H.R. 3893, was passed by the House on October 7 by a vote of 212-210. A similar bill, S. 1772, was defeated in the Environment and Public Works Committee on October 26 by a tie vote of 9-9.

The budget reconciliation process was the vehicle for two major energy initiatives: the opening of part of the Arctic National Wildlife Refuge (ANWR) to oil and gas development and the lifting of the moratorium on oil and gas leasing on much of the Outer Continental Shelf (OCS). Both the Senate Energy and Natural Resources Committee and the House Resources Committee included leasing ANWR in their reconciliation bill sections. The House Resources Committee included provisions regarding OCS leasing, but the Senate Energy Committee did not. ANWR leasing was included in the Senate bill that passed on November 3 (S. 1932), but the House leadership dropped both ANWR and OCS provisions before its

* Excerpted from CRS Report IB10134, dated June 2, 2006.

reconciliation bill, H.R. 4241, was passed on November 18 by a vote of 217-215. The conference report on the bill, approved by the House but amended in the Senate, does not contain either measure.

The gasoline price surge influenced the debate over P.L. 109-58, but the urgency of previous energy crises was lacking. In part, this may be due to the fact that there has been no physical shortage of gasoline or lines at the pump. In addition, the expectation of former crises — that prices were destined to grow ever higher — has not been prevalent.

However, the persistence of high gasoline and oil prices into a second summer has raised alarms over the economic consequences of the situation, heightened following the disastrous effects of Hurricane Katrina.

As gasoline prices surged to the $3.00 per gallon level again in spring 2006, a new wave of legislative proposals was put forth in Congress, including strengthening measures regarding price gouging and profiteering, temporarily refunding federal gasoline taxes, and windfall profits tax measures.

Most Recent Developments

Gasoline prices surged toward $3.00 per gallon in April and stayed there during the first half of May. The approach of the summer driving season, and complications involving the phasing out of the gasoline additive MTBE, were cited by some observers as behind the latest runup.

The price move stimulated a large number of new legislative initiatives and renewed interest in several that had been under consideration for some time. On May 3, the House passed the Federal Energy Price Protection Act (H.R. 5253) to prevent price gouging.

The tentative schedule for the week of June 5 included possible House floor action on H.R. 5254, the Refinery Permit Process Schedule Act of 2006, and other energy legislation.

Background and Analysis

Legislative Activities

The continuing crisis of high gasoline prices has led to a broad spectrum of proposed new legislation. Despite passage of the major Energy Policy Act of 2005 (P.L. 109-58) last summer, many Members are exploring a variety of measures to increase supply and reduce demand in the short term, and to reduce the impact of high prices on consumers. Some are also proposing to revisit longer-term policies, some of which were passed up in the process of reaching agreement on P.L. 109-58.

This issue brief reviews the major legislative initiatives to deal with the gasoline price issue. To put these proposals in perspective, it first describes some of the factors that have led to the high prices of both crude oil and gasoline.

Why Are Prices So High?

The run-up of gasoline prices that began in spring 2004 (see Figure 1) climaxed a period of almost five years during which gasoline prices demonstrated a great deal of regional volatility but less of an increase at the national level. In 2004 a large number of factors combined to exert pressure on gasoline prices in all parts of the country. Some of these factors have affected the price of crude oil, and others the cost of producing and marketing gasoline.

Crude Oil Prices

Past energy crises have demonstrated that oil is traded in a world market, in which events in remote areas affect the price of crude for almost everyone. In the 12-18 months leading up to the crisis, these events included the following:

- Decisions by the Organization of Petroleum Exporting Countries (OPEC) cartel, after having reduced production quotas in 2002, to raise them only slowly and reluctantly;
- Unexpected demand growth in China;
- Disruptions in oil production in major exporters, including Venezuela, Iraq and Nigeria;
- Decline in the value of the U.S. dollar, the currency in which oil is traded in the world market, compared to other major currencies, particularly the Euro.
- Uncertainty and fear of major disruptions in Iraq and Saudi Arabia, in the context of the war in Iraq and the threat of terrorism.

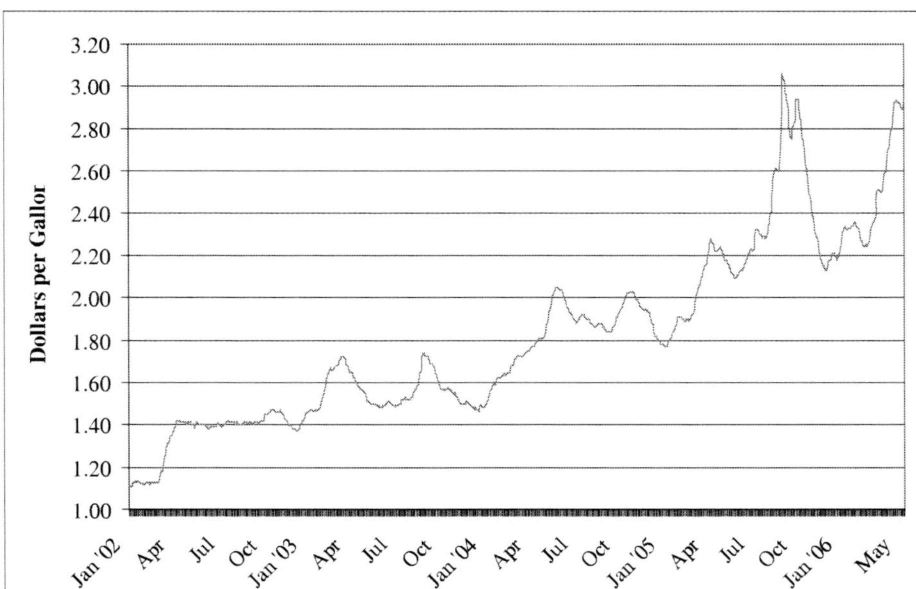

Note: Prices include federal, state and local taxes. Last date above is May 15, 2006.
Source: *Daily Fuel Gauge Report*, American Automobile Association, *[http://www.fuel gaugereport.com]*, compiled by CRS.

Figure 1. Average Daily Nationwide Price of Unleaded Gasoline, January 2002 - May 2006.

How Oil and Gasoline Prices Are Determined

The price of crude oil is set through the interaction of world demand and supply. World demand, at 83.6 million barrels per day in 2005, grew by 4% in 2004, moderated to 1.3% in 2005, and is forecasted to grow by 1.8% in 2006.[1] World supply, at 84.5 million barrels per day, was less than 1 million barrels per day more than demand, leaving relatively little excess supply to draw on if the market was disrupted by natural or political disasters. When excess supply on the market is low, prices tend to rise and become more volatile.

Higher prices for crude oil translate directly into higher prices for gasoline. Crude oil accounts for about 55% of the cost of gasoline. Refining, distributing, and marketing account for about 26% of the cost of gasoline, and taxes account for about 19%.[2] Whether the crude oil a refiner processes is purchased on the open market, or is produced by the oil company itself, higher costs for any element in the cost of gasoline are likely to be passed on to consumers. Gasoline prices have also increased due to limited refining capacity in the United States, the range of fuel blends required to meet air pollution requirements, and the mandated use of ethanol as an additive.

As often happens when commodity prices are volatile, speculation in futures contracts accentuated the upward price pressure and appeared to continue high prices longer than would be expected as market fundamentals push toward lower prices.

Gasoline Prices

Just as a number of factors led to increased crude prices, a combination of features in the U.S. refinery industry contributed to an increase in gasoline prices.

- U.S. demand for gasoline has increased as economic growth has continued (See Figure 2).
- Domestic refining capacity has declined, both in number of refineries —from 324 in 1981 to 153 in 2002 — and in total capacity — from 18.62 million barrels per day (mbd) in 1981 to 16.78 mbd in 2002.
- The structure of the refining industry has changed. In 1981 most refining capacity was owned and operated by integrated oil companies that supplied their own crude oil, refined it, distributed it, and marketed the products. Refining was only one part of the company's profit-making operation, and frequently was not an important profit maker. Now the refining industry is characterized more by independently owned, nonintegrated firms. When refineries are the sole source of revenue to the owners, it becomes more important that the operation be profitable, leading to pressure to raise prices.
- The refining industry has been operating with lower inventories of both crude oil and gasoline, as a means of cutting costs. The side effect has been reduced ability to meet unanticipated demand, leading to greater price pressure.
- Gasoline markets are fragmented regionally because air quality requirements have led to numerous different formulations to meet varying standards. In meeting demand for these regional formulations, called "boutique fuels," refiners lose flexibility to meet local variations in demand elsewhere, leading to increased price pressure.
- With domestic refining capacity constraints, a greater proportion of gasoline demand is being met with imported products. Foreign refiners typically manufacture products designed to sell in the international market, not the special product "boutique fuels" demanded by a significant share of the U.S. market.
- Refiners have had increased costs in the past year to comply with new requirements to limit sulfur content and to switch from the oxygenate additive MTBE to ethanol.

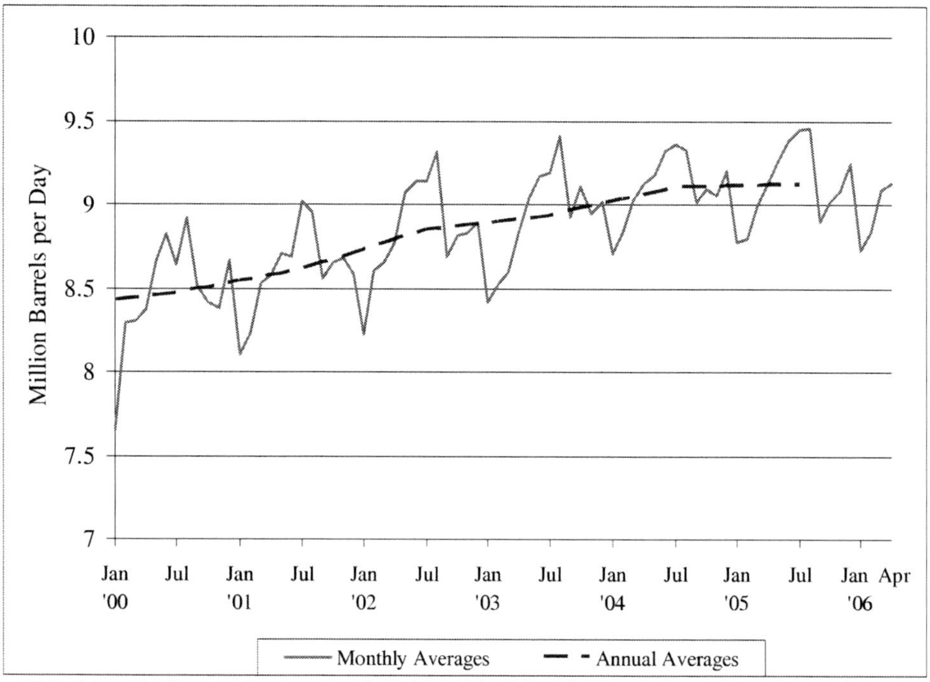

Source: EIA, *Monthly Energy Review*, April 2006, Table 3.4 and *Weekly Petroleum Status Report*, May 10, 2006, Table 10.

Figure 2. Consumption of Motor Gasoline, 2000 - April 2006.

These various factors pushed the nationwide average price of gasoline over $2 per gallon in May 2004. By mid-June, Energy Information Administrator Guy Caruso was able to note a slight decline in prices, and tell a Senate Energy Committee hearing that, "absent major disruptions, oil and gasoline markets may be turning a corner."[3] However, persistent high crude prices pushed gasoline prices over $2 again in October, and yet again in March 2005. By April 2005, Caruso was suggesting that increasing world demand for oil might keep the price of crude above $50 per barrel through 2006.[4] In fact, in August the price surged close to $70.

The price surge intensified discussion of energy policy and led to further calls for passage of energy legislation. However, until the climax of the Katrina disaster, the urgency of previous energy crises has been lacking. Throughout the period, U.S. gasoline consumption continued to rise, although the usual summer peak in consumption appears to have been somewhat blunted in 2004, as shown in Figure 2. In part this may be because, although the price of gasoline in nominal terms set a record, in real terms it did not appear to be reaching the level of the Iranian crisis years of the early 1980s (see Figure 3), again until Katrina pushed it toward the $3.00-per-gallon mark. Further, unlike the earlier crises, there was no physical shortage of gasoline, and no lines at the pump, except in local disaster-affected areas. Consumption of gasoline fell sharply after prices peaked. The drop was typical of the post-Labor Day decline but more steep than in previous years.

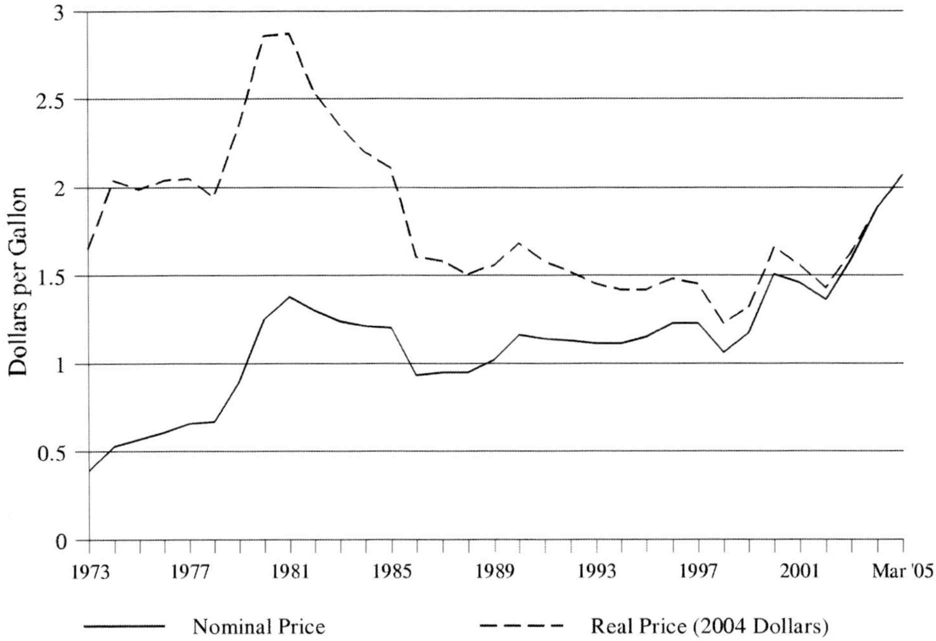

Source: EIA, *Monthly Energy Review*, May 2005, Tables 1.6 and 9.4, calculated by CRS.

Figure 3. Nominal and Real Price of Gasoline, 1973-2004 and March 2005.

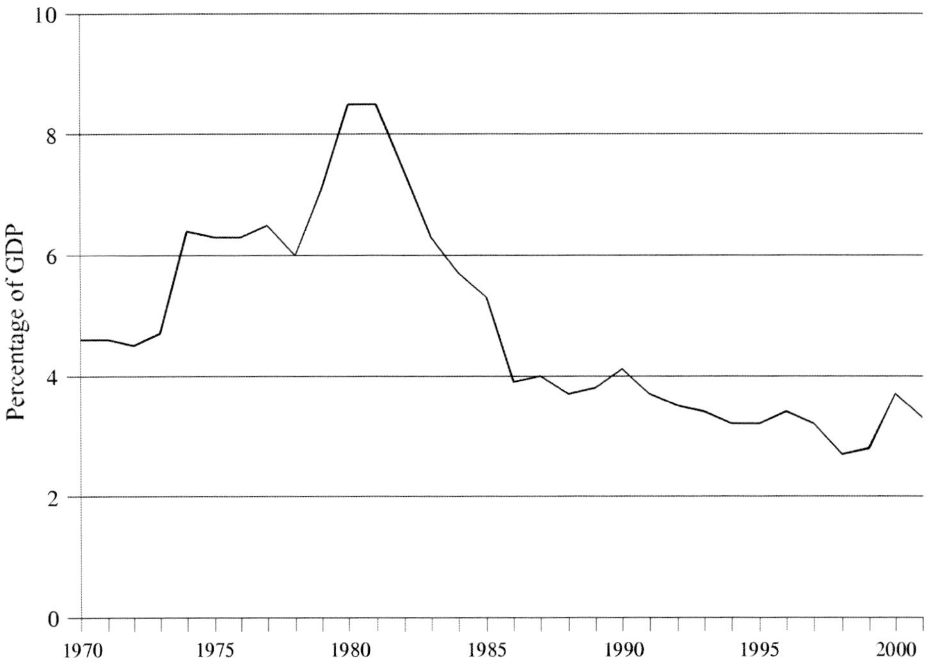

Source: Calculated by CRS with data from EIA, *Annual Energy Review* 2005, Table 3.5. GDP from Bureau of Economic Analysis, Department of Commerce.

Figure 4. Consumer Spending on Oil as % of GDP, 1970 – 2001.

As Figure 4 indicates, the proportion of consumer expenditures on oil and gasoline had declined from the high levels of the 1970s and early 1980s. Data are not yet available to indicate what effect the price run-up starting in 2004 has had on this measure. Perhaps most important, the common view during the earlier crises was that oil prices not only were high, but were destined to become ever higher in the coming years. This view is no longer prevalent, and the general expectation has been that the price increase is a temporary phenomenon, although lasting longer than expected. The current crisis has led to some analytical speculation that world oil production has peaked, but additions to proved world oil reserves seem to contradict that thesis. Oil industry analysts appear confident of a long life remaining for the resource and argue that if oil is replaced, it will be because of improved alternative technologies, not because the world is running out of oil.

As shown in Figure 5, gasoline prices historically have increased less than the general rate of inflation, as measured by the Consumer Price Index (CPI). After the surge in 1973, and again after the 1979-1980 run-up, gasoline prices grew very slowly and even declined, dropping sharply in 1986. A sudden increase in 2000 was similarly followed by slow or declining prices. During the current run-up, gasoline price increases have far outpaced the general CPI increase.

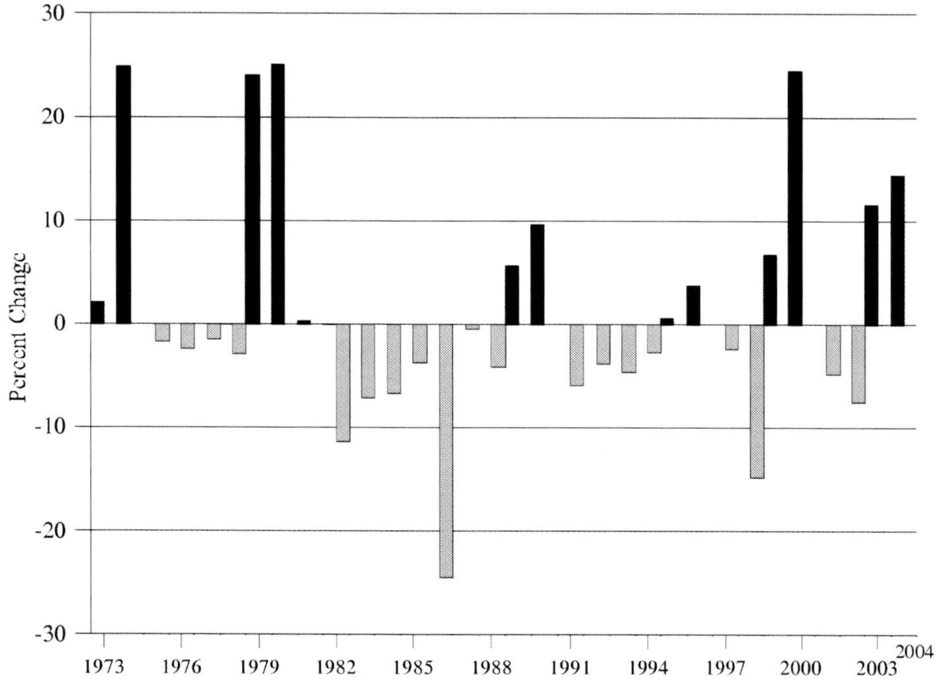

Source: EIA, *Annual Energy Review* 2004, Table 5.24. CPI from the Bureau of Labor Statistics. Calculated by CRS.

Figure 5. Percent Change in Gasoline Prices Compared to the Consumer Price Index, 1973-2004.

Policy Options

The several energy crises of the past led to major legislative action, twice in the 1970s and once following the 1991 Gulf War. The just-passed Energy Policy Act of 2005 differs from the previous actions because the Congress had been considering major energy legislation for three years before the situation became a nationwide concern. By the time the bill finally moved through the Congress, the major issues had already been fully debated, and the final version differed little from previous initiatives except for resolving a number of issues that had blocked passage before.

As in previous legislative energy debates, a major policy divide existed between those who view the gasoline-fueled automobile as a temporary necessity to be tolerated only until a substitute fuel or alternative means of transportation can be developed, and those who expect oil to be the same dominant transportation fuel in the indefinite future that it is at present. Compromise agreements have been reached via a combination of measures that enhance the development of alternatives or restrain the growth in demand for oil, on the one hand, and those that increase production or reduce the cost of supplying that demand, on the other. However, individual measures often carry with them complicating features that make passage more difficult. In addition, major legislation often becomes the vehicle for measures that typically would not find enough support to pass as individual bills, or which may be added to gain support for the whole measure. In the legislative climate of the 108th Congress, balancing the various interests involved proved too difficult a task, despite the influence of a nationwide energy crisis in an election year. Under the stimulation of continued high oil and gasoline prices, the 109th Congress pursued the goal again, this time successfully. However, as gasoline prices continued to surge, and damage to Gulf of Mexico oil and gas resources and facilities by Hurricane Katrina was assessed, calls for further measures to address the crisis were heard in Congress. Continued price pressure into the spring of 2006 has solidified the calls into specific legislative proposals.

Oil-Related Legislation

Reducing Impacts on Consumers

A number of proposals are aimed at easing the impact of high prices on consumers, or are aimed at the oil industry's price-making policies.

Price Gouging

The rapid increase in gasoline prices following the Katrina disaster led to allegations of price gouging. P.L. 109-58 included a provision requiring the Federal Trade Commission (FTC) to conduct an investigation into price gouging in the recent increases in gasoline prices. H.R. 3893, as passed the House October 7, includes provisions requiring FTC to define price gouging and penalize violators.

After gasoline prices surged again in April 2006, momentum increased for various price-gouging proposals. Several initiatives that would impose windfall profits taxes on crude oil profits were also introduced. On May 3, the House passed the Federal Energy Price Protection

Act, H.R. 5253, by a vote of 389-34. The bill would require the Federal Trade Commission within six months after passage to issue rules defining price gouging in crude oil, gasoline, diesel, home heating oil, and biofuel, and would prohibit gouging so defined.

Tax Relief

On April 26, 2006, Senator Domenici proposed an amendment to the FY2006 emergency supplemental appropriations bill (H.R. 4939) that included a $100 "Fuel Tax Holiday Rebate." The proposal was cosponsored by Majority Leader Frist and other Senators. The amendment also contained provisions about price-gouging, easing restrictions on rebates for hybrid vehicles, and approving leasing of ANWR. However, the amendment was withdrawn May 2.

Suspending Import Duties on Ethanol

Because the latest gasoline price run-up was attributed in part to the need to increase use of ethanol, some Members, including Senate Foreign Relations Committee Chairman Lugar and House Energy and Commerce Chairman Barton, have proposed suspending temporarily duties on imported ethanol, particularly from Brazil.

"Boutique" Fuels

The Energy Policy Act of 2005 had some provisions related to local requirements for specific blends of gasoline, but some proposed legislation is aimed at further reducing "boutique fuels" requirements that, according to some observers, make the national gasoline market less flexible. H.R. 3893, the Gasoline for America's Security Act of 2005, passed by the House October 7, 2005, by a vote of 212-210, would among other provisions require EPA to develop a Federal Fuels List and to limit local gasoline blends to those on the list. (For details, see CRS Report RL31361, *"Boutique Fuels" and Reformulated Gasoline: Harmonization of Fuel Standards*, by Brent D. Yacobucci.) Other boutique fuels legislation has also been proposed.

Windfall Profits Tax

Several bills introduced would impose taxes on oil companies in light of record profits recorded as a result of the crude and gasoline price run-up. The context of the proposals lies in the Crude Oil Windfall Profit Tax Act (P.L. 96-223) of 1980, as part of a compromise between the Carter Administration and the Congress over the decontrol of crude oil prices.

Mid- To Long-Term Supply and Demand

Most proposals affecting supply and demand of crude oil and gasoline would not affect the current short-term crisis but would be aimed at longer term trends.

CAFE

Automobile fuel economy standards also have a long history of controversy, going back to their establishment in the 1970s. In the mid-1990s, the National Highway Traffic Safety

Administration (NHTSA) was considering a rulemaking that would result in increased standards for light-duty trucks (including sport utility vehicles), but for several years, Congress included in its annual appropriation for NHTSA a measure prohibiting NHTSA from analyzing or undertaking such a ruling. That prohibition was dropped in the FY2004 NHTSA appropriations, and a final rule issued by NHTSA in April 2003 requires a boost in light-truck fuel economy to 22.2 miles per gallon by model year 2007. In the summer of 2005, the Bush administration proposed new fuel economy standards for light trucks, to take effect in the 2008 model year.

During House floor debate on P.L. 109-58, an amendment to increase fuel economy standards to 33 miles per gallon over 10 years was defeated by a vote of 177-254. A more general amendment to the House bill, requiring the Administration to take "voluntary, regulatory, and other actions" to reduce oil demand in the United States by 1 million barrels per day from projected levels by 2013 was defeated 166-262. The measure was included in the bill passed by the Senate, but was dropped in conference.

Continued high gasoline prices have raised congressional interest in higher mandated CAFE standards again. Among the proposals is H.R. 3762, which would require average fuel economy of 33 miles per gallon by 2016. Other proposals would give NHTSA authority to raise standards above the present level and would change the vehicle classifications to one based on weight. On May 10, the House Energy and Commerce Committee ordered reported a bill that would authorize the Secretary of Transportation to set fuel economy standards for passenger automobiles based on one or more vehicle attributes.

ANWR

Oil and gas exploration and development of part of the Arctic National Wildlife Refuge (ANWR) have been controversial for many years. This was part of the early proposals for legislation that eventually became the Energy Policy Act of 1992, but was dropped in the face of strong opposition in both houses. Support for action grew gradually through the decade, along with technological developments that advocates claimed would reduce the environmental impact of development; the House included a development measure in its version of an omnibus energy bill in August 2001. A similar measure was part of the House-passed legislation in the 108th Congress. Opposition in the Senate kept the measure from the floor, however, and it was dropped in conference.

In the 109th Congress, Senate supporters of ANWR development moved the issue to the budget process, where it can be approved by a simple majority vote. On March 9, 2005, the Senate Budget Committee issued a FY2006 budget resolution that assumes $2.4 billion of revenue over five years from leases in ANWR. On March 16 the Senate rejected an amendment by Senator Cantwell to strike the ANWR provisions, by a vote of 49-51. The next day the Senate passed the budget resolution (S.Con.Res. 18). The measure was included in the package of provisions for the budget reconciliation bill approved by both the Senate Energy Committee and the House Resources Committee. However, the House leadership removed the ANWR provisions from its reconciliation bill (H.R. 4241) before it was passed on November 18 by a vote of 217-215.

Despite the failure of the ANWR provision, the Administration's FY2007 budget request for the Department of the Interior, submitted February 6, assumes that $8 billion can be raised in 2008 through leasing of oil and gas resources in ANWR.

Savings Goals

A number of legislative proposals would set goals for reducing oil consumption. An example is the Enhanced Energy Security Act of 2006 (S. 2747), introduced by Senator Bingaman May 4, 2006, which would require the Director of the Office of Management and Budget to develop an action plan to save 2.5 mbd in 2016, 7 mbd in 2026, and 10 mbd in 2031.

OCS Leasing

The moratorium on oil and gas leasing in the Outer Continental Shelf (OCS), except in the central and western Gulf of Mexico and some parts of Alaska, was subject to much controversy during consideration of P.L. 109-58. A proposal to allow states to voluntarily opt out of the moratorium was dropped under threat of filibuster, and even a measure to order the Department of the Interior to perform an inventory of OCS resources barely survived the debate.

Following the disruption of production by Katrina, momentum to lift the moratorium increased, and some supporters suggested it might be included in the budget reconciliation process. The House Resources Committee on September 28 marked up a bill that would have given states the option of allowing drilling for petroleum and natural gas. In approving the bill, the committee adopted an amendment that would lift the moratoriums on drilling the OCS for natural gas completely. In the face of opposition to the natural gas provision, Representative Pombo, Chairman of the Resources Committee, decided not to bring the bill to the floor as planned.

On October 26, the Resources Committee, as part of its package of measures for the budget reconciliation bill, included provisions that would make statutory the current presidential moratorium on OCS leasing until 2012 but would allow individual states to opt out of the moratorium and would allow states close to 50% of the royalties from oil and gas production that resulted. It would also give states the option to extend the moratorium after 2012. However, the provision was removed from the bill (H.R. 4241) before it was passed on November 18 by a vote of 217-215.

On May 10, 2006, the House Appropriations Subcommittee on Interior and the Environment voted to lift the congressional suspension of natural gas leasing for much of the OCS. The legislation, if approved, would leave in place the presidential moratorium on the same region, which operates until 2012.

Refinery Revitalization

P.L. 109-58 contained some provisions to encourage construction of new oil refineries, but the destruction to refining facilities caused by Katrina in the Gulf of Mexico area led to calls for further measures. On September 28, 2005, the House Energy and Commerce Committee reported out H.R. 3893, the Gasoline for America's Security Act of 2005, and the House passed the bill October 7 by a vote of 212-210. Among other measures, the bill would provide for presidential designation of potential refinery sites on federal lands and military bases that are closing, and set up a process for coordinating authorization and related environmental reviews for construction of new refineries, to be led by the Department of Energy. It would centralize judicial review of the process in the U.S. Court of Appeals for the District of

Columbia. The bill as reported would also have amended the Clean Air Act regarding provisions for New Source Review for refineries and other facilities, but the measure was dropped before being brought to the House floor. (For details, see CRS Report RS21608, *Clean Air and New Source Review: Defining Routine Maintenance,* by Larry Parker.) The bill also contains measures concerning boutique fuels (see above).

A similar bill was introduced in the Senate September 26, as the Gas Petroleum Refiner Improvement and Community Empowerment Act, S. 1772, but the bill was rejected October 26 by the Environment and Public Works Committee by a vote of 9-9.

On May 3, 2006, the House brought up under suspension of the rules the Refinery Permit Process Schedule Act (H.R. 5254), but the 237-188 vote was less than the two-thirds approval required under suspension. The bill, which would require the President to appoint a "federal coordinator" to organize the permitting of new refineries, is expected to come to the floor under regular rules the week of June 8.

Presidential Proposals: The Advanced Energy Initiative

In his January 31 State of the Union message, President Bush set the goal of breaking the U.S. "addiction to foreign oil" and of "replacing" more than 75% of oil imports from the Middle East by 2025. In specifying the Middle East as the source, the proposal differed from provisions considered in the previous year to set a goal of reducing total imports.

The main thrust of the presidential initiative is to increase funding for research in producing ethanol from plant fiber biomass (rather than from corn), for improved batteries for hybrid automobiles, and for hydrogen fuels. At the same time, the budget request for FY2007 for DOE, which includes these increases, would eliminate programs for oil and gas research.

Legislation

109th Congress

P.L. 109-58, H.R. 6

Energy Policy Act of 2005. Introduced April 18, 2005. Passed by the House April 21, 2005 (249-183). Passed the Senate June 28. Reported out of conference July 26 and passed by both houses July 28 and July 29. Signed into law August 8, 2005.

H.R. 3893 (Barton)

The Gasoline for America's Security Act of 2005. Introduced September 26, 2005. Reported out by the House Energy and Commerce Committee September 28. Passed by the House October 7 by a vote of 212-210.

H.R. 5253 (Barton)

The Federal Energy Price Protection Act of 2006. Passed by the House May 3, 2006, under suspension of the rules by a vote of 389-34.

H.R. 5254 (Barton)

The Refinery Permit Process Schedule Act of 2006. Rejected by the House under Suspension of the Rules May 3, 2006, by a vote of 237 yeas to 188 nays.

H.R. 3762 (Boehlert)

To require higher standards of automobile fuel efficiency in order to reduce the amount of oil used for fuel by automobiles in the United States by 10% beginning in 2016, and for other purposes.

S. 1772 (Inhofe)

The Gas Petroleum Refiner Improvement and Community Empowerment Act. Introduced September 26, 2005. Rejected by the Senate Environment and Public Works Committee October 26 by a vote of 9-9.

S. 555 (DeWine)

The No Oil Producing and Exporting Cartel (NOPEC) Act of 2005. Introduced March 8, 2005. Adopted by voice vote as an amendment to H.R. 6 in the Senate, June 21. Not included in the final version of H.R. 6.

S. 2747 (Bingaman)

The Enhanced Energy Security Act of 2006. Introduced May 4, 2006.

S. 2748 (Bingaman)

The Enhanced Energy Security Tax Incentives Act of 2006. Introduced May 4, 2006.

References

[1] International Energy Agency, *Oil Market Report,* April 12, 2006. pp.4-6.
[2] Energy Information Administration data based on March 2006 data and a base price of gasoline of $2.43 per gallon. See [http://www.eia.doe.gov].
[3] Guy Caruso, statement before the Senate Committee on Energy and Natural Resources, June 15, 2004.
[4] "EIA: Oil to remain above $50 through '06." *Oil Daily*, April 8, 2005, p. 1. Report of a speech by Guy Caruso at the National Press Club in Washington, DC, April 7, 2005.

In: New Research on Energy Economics
Editors: P.G. Caldwell and E.V. Taylor, pp. 211-217
ISBN: 978-1-60456-354-2

Chapter 9

ENERGY, ENVIRONMENT AND POVERTY NEXUS IN DEVELOPING COUNTRIES*

Dhulasi Birundha Varadarajan and K. Arockia Raj
Department of Environmental Economics, School of Economics
Madurai Kamaraj University, Madurai -625 021, South India

Introduction

A growing consciousness of the intertwined issues of environment and development and the particular role of energy in the nexus of these issues have been realised at present. As a result of this awareness, voices of concern in support of Environmentally Sound and Sustainable Development (ESSD) are now increasingly heard. The great challenge now is to move from the voicing of concerns to accelerated action and change. The list of environmental issues that are relevant to any discussion of ESSD strategies is long and diverse. Implementing development strategies which include the alleviation of poverty as a central focus must address both the direct and indirect linkages between these environmental issues and poverty.

The Dimensions of Poverty

The concept of poverty refers to an individual's (or family's) lack of access, associated primarily with inadequate income, to basic human needs such as food, shelter, clothing, safe water, sanitation, healthcare and education.

Operationally, however, poverty standards are typically expressed in single dimensions, the monetary resources that would enable an individual to consume either a fixed bundle of basic goods and services (absolute poverty) or some fraction of the bundle of goods and services that some reference group is able to or actually does, consume (relative poverty).

It has been estimated that, as of 1993, roughly 1.3 billion people in developing countries – 30 percent of their total population – consume less than one dollar per day worth of goods

* Originally printed in Readings in World Development: Growth and Development in the Asia Pacific, edited by K.C. Roy and S. Chatterjee, Nova Science Publishers, 2006, ISBN 1-59454-947-8

and services. Statistics on the inability of people in developing countries to satisfy basic human needs corroborate the enormous scale of poverty and highlight its breadth and complexity. For example, an estimated 20 percent of people in developing countries do not have access to health services; 30 percent lack access to safe water and 61 percent lack access to sanitation.

Poverty and Environment

The linkages between poverty and the environment are far too complex. An argument is generally made that poor people have a tendency of overusing resources like land, forests and water, thereby degrading them. It was even said one time that poverty is the greatest polluter. There is some evidence to say that the poor are more dependent upon such natural resources. The number of poor being large in several countries, the total environmental damage may look too high. Even then, it is the type of property rights (such as open access, or neglected state properties) that can be a major cause of environmental degradation.

The neglect of the environment in development economics is ironic, because people in poorcountries are in great part agrarian and pastoral. Rural people account for about 65 per cent of the population of what the World Bank classifies as low income countries. The proportion of the total labour force in agriculture is a bit in excess of this. The share of agriculture in gross domestic product in these countries is 30 percent. These figures should be contrasted with those from industrial market economies, which are 6 percent and 2 percent, respectively, for the latter two indices (Partha Das Gupta, 1996).

Poor countries are in large measures biomass based subsistence economies, in that the rural poor eke out a living from products obtained directly from their local environment. For example, in its informative study of life in a micro water shed of the Alaknanda river in the Central Himalayas in India, the Indian Centre for Science and Environment reports that, of the total number of hours worked by the villagers sampled, 30 percent was devoted to cultivation, 20 percent to fodder collection, and about 25 percent spread evenly between fuel collection, animal care and grazing. Some 20 percent of time was spent on household chores, of which cooking took up the greatest portion and the remaining 5 percent involved other activities such as marketing.

In Central and West Africa, it has been shown how vital forest products are to the lives of rural people. In many instances, natural resources are of direct value to us as needs for sustaining our lives or as consumption goods (e.g., breathable air, drinkable water and fisheries), in others they are of indirect value, sometimes they are both. The value may be utilitarian (e.g., the resource may be a source of food, or a key stone species in an ecosystem), it may be aesthetic (e.g., a landscape) or it may be intrinsic (e.g., it could be a living animal), indeed, it may be all these things at once. Resource stocks are measured in different ways depending on their character: in mass units (e.g., biomass units for forest, cow-dung and crop residues in number e.g., size of animal herd); in indices of quality e.g., water and air-quality indicators in volume units (e.g., acre-feet for aquifers) and so forth.

Problems of poverty and environmental degradation in developing countries are closely linked (WCED, 1987). The majority of poor people live in rural areas and derive most of their income from soil and forest resources (World Bank, 1990). It has been claimed that poverty may lead to short-planning horizons, which may prevent poor farm households from investing in conservation to protect their natural base. Yet, there have been very few empirical studies of the planning horizon or Rates of Time Preferences (RTPS) among rural poor.

Environmental degradation such as eroding soil, receding forests and vanishing water supplies, is a cause of accentuated poverty among the rural poor in poor countries. There is truth in this. But there is also much accumulated evidence that poverty itself can be a cause of environmental degradation.

This reverse causality arises because some environmental resources (e.g., ponds and rivers) are essential for survival in normal times, while others are a source of supplementary income in times of active economic stress. This mutual influence can offer a pathway along which poverty, environmental degradation and even high fertility feed upon one another in a synergistic manner over time. Indeed, an erosion of the environmental resource base can make certain categories of people destitute even while the economy's GNP increases.

Poverty Environment and Development

Based on the studies after Kuznetz, it can now be said that there is a clear inverse relationship between increase in poverty and economic growth. The relationship between economic development and environmental degradation is quite mixed, between U-shaped, inverted U-shaped, or S-shaped etc. The major issues in the poverty environment and developments are:

An increase in poverty within the community results in increased environmental degradation. Conversely, a reduction in poverty will increase environmental degradation. Other things being equal, an environment inhabited by the poor will be more degraded and become degraded than one inhabited by the rich.

There have been important social changes (policies, external events) that have resulted in concurrent increases in poverty and environmental degradation in a number of developing countries.

Deterioration of the ambient environment hurts the poor more than the rich. But, Jaganathan (1989) looked at rates of deforestation and the level of poverty in West Java and land use and poverty in Nigeria and found little evidence that poverty was a driving force in the deforestation or in the damaging changes in land use.

Policies that change the environment can hurt the poor more than the rich and vice versa. In this regard, Linde Rahr (1998) analyses the effects of income and gender on tree planting. He found that tree planting increases with the increase in female numbers of household members and also increases with the increase in male income.

CPRs as Source of Livelihood

Common property resources also provide the rural poor with partial protection in times of unusual economic stress. For landless people they may be the only non-human assets at their disposal. A number of resources, such as fuel, wood and water for home use, berries and nuts, medicinal herbs, resin and gums are the responsibility of women and children.

The extent of common-property resources as a proportion of total assets in a community varies considerably across ecological zones. In India, they appear to be most prominent in arid regions, mountain regions, and un-irrigated areas. They are least prominent in humid regions and river valleys. The links between under nourishment, destitution and an erosion of the rural common property resource base are close (Das Gupta, 1993, 1996). Jodha (1986) used data from over eighty villages in twelve, districts from six dry tropical states in India to estimate that, among poor families, the proportion of income based directly on the local commons is for the most part in the range 15-25 percent (Jodha, 1995). In his work on the dry lands of India, Jodha (1986) noted a decline in the geographical area covering common property resources ranging from 26 to 63 percent over a twenty-year period. This was in part due to the privatization of land, a good deal of which in his sample had been awarded to the rural non-poor.

The impact of these forms of privatization on the poorest of the poor is confirmed by economic theory. They suggest that privatization of village commons and forest lands, while hallowed at the altar of economic efficiency, can have disastrous distributional consequences, disenfranchising entire classes of people from economic citizenship.

Energy Poverty Nexus

Energy services are a crucial input to the primary development challenge of providing adequate food, shelter, clothing, water, sanitation medical care, schooling and access to information. Energy therefore is one dimension or determinant of poverty and development, but it is vital.

Energy supports the provision of basic needs such as cooking food, a comfortable living, temperature, high living, the use of appliances, pumping of piped water or sewerage, essential health care, educational aids, communication. Energy also fuels productive activities, including agriculture, commerce, manufacturing, industry and mining. Conversely, a lack of access to energy contributes to economic decline. The energy dimension of poverty – energy poverty – may be defined as the absence of sufficient choice in accessing adequate, affordable, reliable, quality, safe and environmentally benign energy services to support economic and human development.

Biomass Consumption

The reality is that biomass energy is the fuel for the poor. Biomass fuels are used by 3 billion, or half of the Global population, to cook their food everyday. Biomass is contributing as a source of energy in the development and developing world. South Asia, with a population of over 1.2 billion, is largely dependent on biomass energy for meeting the energy needs in member countries. Contribution of such traditional biomass energy in the total energy consumption varies from 33 percent in India to 92 percent in Nepal (Pradeep Chaturvedi, 2000). The WEC projections of biomass energy use are given in Table 1, which shows increasing dependence of biomass energy use.

FAO's forestry department recently established a Wood Energy Information System (WEIS) which collects the existing data through the implementation of Regional studies (Wood Energy Today for Tomorrow (WETT), 1977). Table 2 explains the wood fuel consumption from 147 countries arranged in line with the FAO classification already used in other documents, such as the state of the world forests. The total production of wood in 1995 reached approx 3900 million cum, out of which 2300 million cum have been used for wood fuels (Table 2). This means that approximately 60 percent of the world's total wood removals from forest and non-forest land have been used for energy purposes. FAO's Regional Wood Energy Development Programme for Asia operating in 16 developing countries in Asia, collected data for their member countries and projected that two-thirds of wood fuels come from non-forest resources (1997).

World production of fuel wood for energy use in 1995 is estimated to be about 1.9 billion cum or 1.4 billion tonnes (FAO, 1998). An estimated additional amount equivalent to some 0.3 billion tonnes of wood is recovered and recycled for energy use. The total of 1.7 billion tonnes of fuel wood is equivalent to 550 million tonnes of oil. Compared to 9 giga tonnes of oil equivalent of primary energy supply in 1990, it stands at 6.7 percent. Most of it (76 percent) is used in developing countries. The balance of 24 percent fuel wood used in developed countries represents only 2 percent of their total energy consumption; 44 percent of the total wood fuel consumed in the world is used in Asian countries.

Table 1. WEC Projection of Bio-Mass Energy use 1998 Role of Bio-mass in Future Global Energy Use (EJ)

Scenario	Year	2025	2050	2100
Shell	1996	85	200-220	-
IPCC	1996	72	280	320
Green Peace	1993	114	181	-
Johnsson et al	1993	145	206	-
WEC	1993	59	94-157	132-215
Dessus et al	1992	135	-	-
Lashof and Tirpak	1991	10	215	-

Source: Hall and Screase – WEC's, World Resource Handbook (1998)

Table 2. Total Wood Fuel Consumption Regions

	1000 cum	PJ	Ratio wood Fuel / All Energy	Ratio Wood Fuel / All Wood Uses
DevelopmentCountries	1763262	17633	0.15	0.80
Developed Countries	536754	5368	0.02	0.31
World	2300016	23000	0.07	0.59

Source: FAO State, United Nations Energy Year Book 1997, Wood Energy Today for Tomorrow Studies (FAO 1997)

Energy Carriers

Roughly 2 billion people depend mainly on traditional fuels for cooking. Households use fuel in a variety of activities, including cooking, water heating and lighting. Different energy carriers can be used for each of these activities. These carriers form what is commonly referred to as an energy ladder for that activity. It is well known that people living in poverty tend to rely on a significantly different set of energy carriers than the rich.

Each rung corresponds to the dominant fuel used by a particular income group, and occupy different rungs (Hozier and Dowd, 1987; Reddy and Reddy, 1994). Wood, dung and other biomass fuels represent the lowest rung on the energy ladder for cooking. Charcoal and coal (when available) and kerosene represent higher steps up the ladder to the highest rungs, electricity and LPG.

Environmental Aspects of Biomass Consumption

The ordering of fuels on the energy ladder also tends to correspond to the efficiency of the associated systems and their cleanliness. For example, the cooking stove efficiencies of firewood, kerosene and gas are roughly 15, 50 and 65 percent respectively.

The emission of wood energy systems normally depend on the type and scale of the system and its operation. Conversion related emissions become lower when the scale of conversion unit increases. The most important emissions that can occur are carbon monoxide Polycylo-Aromatic-Carbon Hydrogens (PACs), Nitrogen Oxide (NO_X) particles. Thermal efficiencies in small scale household wood heaters can be around 50 percent but are much lower in less efficient designs.

It appears that the patterns of energy consumption of people living in poverty tend to increase their misery and aggravate their poverty (Leach, 1992; Das Gupta, 1993) for the following reasons: the use of biomass compromises the health of household members, especially when it is burned indoors without a proper stove to help control the generation of smoke outside. Thus, in addition to its relatively high cost, the use of biomass fuel may also promote higher medical care expenditure and diminish the ability of the poor to work productively.

Traditionally, biomass energy has been associated with environmental degradation and health hazards. Biomass for direct combustion is widely discussed. Production of bio-ethanol for transport vehicles is an emerging application where biomass can substantially avoid carbon emissions. Bio fuel productionregarded as a carbon sink; the total amount of carbon stored per hectare under the bio-energy crop is greater than the level of stored carbon in the vegetation previously on that land.

In Indiathe National Project on Biogas Development (NPBD, 1981-82) aimed at setting up 12 million biogas plants based on cattle dung, and 2.7 million biogas plants have been set up till March 1998. In the year 1997-98, the improved cooking stove on average saved about 400 kg of fuel wood equivalent per year. Fuel wood based biomass gasifier systems, with capacitities of up to 500 Kw, have been developed indigenously and are manufactured in the country. The biomass based cogeneration programme of 25-30 MW has been attracting the interest of a number of small and medium industries from other countries.

Conclusion

To conclude, the challenges of finding more sustainable growth paths and dealing with the nexus of poverty, energy and environment are too great and the intellectual and financial resources available are too scarce for multiple agendas and insensitive or ill-conceived positions.

A consideration of characteristics associated with sustainable development should include: development strategies to provide all people, especially the poor, with livelihoods which are relatively non-polluting, non-resource degrading and less intensive with respect to purchased inputs.

Solving the many problems of the poverty – environment nexus will require many changes, including the empowerment of poor people. Only when poor people have the requisite energy and other natural resources as well as other factors of income and wealth generation, will they be able to and want to act to conserve these resources.

Poor people must have access and control over not just energy and other natural resources, but the other factors of production critical to sustainable development, namely management capacity, to implement this know-how.

Sustainable development means information and management intensive strategies. It is better to educate the rural mass about other matters – namely, the complexities of these intertwined problems of poverty, other natural resource management and the environment, and the challenges in dealing with these problems in societies without well-developed markets, regulatory entities and physical and institutional infrastructure.

Avoid natural resource degradation or depletion except where the economic calculus allowing such depletion explicitly considers the spatial and intertemporal costs and externalities of such depletion. Externalities and use of the common problems are ubiquitous in the developing countries as traditional mechanisms and regulations for common resource management. They often collapse under demographic, commercial and climatic forces.

Thus, if poor people are to have the option of this development path in the midst of many changes, tremendous political, social and financial shifts will be needed. Whether such changes are likely is questionable.

References

Jaganathan, V.N. (1989), *Poverty, Public Policies and the Environment*, Environment Working Paper No.24, Washington DC: The World Bank.

Jodha, N.S. (1986), *Common Property Resources and the rural poor*, Economic and Political Weekly, 21.

Jodha, N.S. (1995), *Common Property Resources and the Environmental context*: Role of biophysical versus social stress, Economic and Political Weekly, 30, 3278-3283.

Linde – Rahr (1998), *Rural Deforestation: Gender Effects on Private Investments in Vietnam*, Working paper, Department of Economics: G. Teborg University.

WCED (1987), *Our Common Future*, Oxford University Press, New York.

World Bank (1990), *World Development Report*, The World Bank: Washington DC.

INDEX

D

E

F

G

H

I

J

K

L

M

N

O

P

Q

R

S

T

U

V

W

Y

Z